# Industrial Boilers

General editor: David Browning

# Industrial Boilers

David Gunn, BSc(Eng), MSc, MIGasE, SFIE, CEng
Robert Horton, BSc (Eng), MIMechE, CEng

Longman
Scientific &
Technical

Copublished in the United States with
John Wiley & Sons, Inc., New York

**Longman Scientific & Technical**
Longman Group UK Limited
Longman House, Burnt Mill, Harlow
Essex CM20 2JE, England
*and Associated Companies throughout the world*

*Copublished in the United States with*
*John Wiley & Sons, Inc., 605 Third Avenue, New York, NY 10158*

*First published 1989*

**British Library Cataloguing in Publication Data**
Gunn, David
   Industrial boilers
   1. Boilers. Design & construction –
   Manuals
   I. Title     II. Horton, Robert
   621.1′84

ISBN 0-582-02532-X

**Library of Congress Cataloging in Publication Data**
Gunn, David, 1912–
   Industrial boilers/David Gunn, Robert Horton.
      p.    cm.
   Includes index.
   ISBN 0-470-21055-9 (Wiley)
   1. Boilers.    I. Horton, Robert, 1929–          II. Title.
   TJ263.5.G86 1988
   621.1′83 – dc19

Set in Linotron 202 10/12pt Times

Printed and bound in Great Britain,
at the Bath Press, Avon.

# Contents

# Preface

The purpose of this book is to describe the evolution of steam and hot-water boilers from the inception to many of the variants now used in industry throughout the world. In doing this we have endeavoured to provide guidance, for both designers and users, on the fundamental subjects of combustion and heat transfer which are essential to the development of efficient and reliable boilers. In this connection we have included consideration of boiler auxiliaries, firing appliances, chimneys, controls and instrumentation. We have covered a wide variety of fuels, and waste-heat boilers. It is our hope that students of technology, manufacturers, and users will find the work to be of benefit to their activities.

As the scope of this book is very wide, it has only been possible to give a broad view of the subject. Numerous references are given for the benefit of those readers who may wish to study specialised areas in greater depth.

The views expressed in this book are entirely those of the authors and it is accepted that there are a number of issues with which readers may disagree. This will perhaps indicate to such readers the wide range of the subject covered.

It is appreciated that there are many designs of boiler available that are not described; it has not been possible to cover all possibilities within the scope of this volume.

Typical design calculations for water-tube boilers have not been given since this would involve the use of proprietary information. They may also give misleading results should the methods be applied out of context on a completely different design of boiler from that for which the data were intended to be used.

While some examples of manual calculations are given, it should be appreciated that the use of computers eliminates the laborious process of carrying out such repetitive calculations. Their use also enables many alternative design aspects to be considered in a short period of time, thus enabling design optimization to be more readily achieved than hitherto. Nevertheless, the authors are of the opinion that design engineers should be able to carry out manual calculations to ensure that they fully understand the basic design principles involved, are able to feed sensible design data into the computers, and are able to interpret computer results into practical boiler designs.

Where reference is made to British Standards and other national codes, reference to the latest editions has usually been made. It should, however, be appreciated that these are continually under review and on occasions are completely revised, therefore specific page and illustration references will only apply to the issue indicated.

# Acknowledgements

The authors wish to thank NEI International Combustion Limited (by which firm Bob Horton is currently and David Gunn was formerly employed) for permission to reproduce many of the illustrations used and to which specific reference is not made in the titles. Our thanks go to the companies specifically mentioned in the titles of illustrations for their kind and patient assistance in providing the illustrations actually used or upon which those produced are based. We wish to acknowledge the help of Mr Peter Wilmot who produced many of the illustrations in a form suitable for publication, of Mr Paul Williamson for reviewing the text and of Mr David Evans of White Young Prentice Royle, Consulting Engineers, for his help in reading the proofs.

We thank Dr Alan Gray of Leeds University for his advice on Chapter 3, and also Ken DeVille and Norman Keighley of NEI for their assistance on Chapters 6 and 8 respectively.

Thanks are also due to a number of ladies for their assistance and patience with the typing of the drafts and final text.

Extracts from BS 2790 : 1986 are reproduced by permission of the British Standards Institution. Complete copies of this and other British Standards can be obtained from the British Standards Institution at Linford Wood, Milton Keynes, MK14 6LE.

Last, but not by any means least, we would like to thank our wives for their patience during the many hours taken to produce this volume.

*D. Gunn*
*R. Horton*

*1988*

# Glossary

**Angle of repose:**   the included angle between the surface of a tipped material which may be granular or fibrous, and the horizontal, after settlement is complete.

**Atomisation:**   the process whereby a liquid is broken down into very small droplets.

**Balanced draught:**   the condition achieved when the pressure of the gas in a furnace is the same as or slightly below that of the atmosphere in the enclosure or building housing it.

**Blowdown:**   the removal of a quantity of water from a boiler in order to achieve an acceptable concentration of dissolved and suspended solids in the boiler water, see Chapter 6.

**Bluff body:**   an obstruction to the flow of gas in a duct whereby eddies are created on the downstream side. In a burner for gas or oil, such a device is often incorporated to promote mixing of the fuel and air and to establish flame stabilisation.

**Calorific value (gross):**   the amount of heat liberated by the complete combustion, under specified conditions, of a unit volume of a gas or of a unit mass of a solid or liquid fuel, in the determination of which the water produced by combustion of the fuel is assumed to be completely condensed and its latent and sensible heat made available. (See BS 526 and Section 3.5.)

**Calorific value (net):**   the amount of heat generated by the complete combustion, under specified conditions, of a unit volume of a gas or of a unit mass of a solid or liquid fuel, in the determination of which the water produced by the combustion of the fuel is assumed to remain as a vapour. (See BS 526 and Section 3.5.)

**Cavitation:**   the release of steam bubbles from hot liquid under conditions of low pressure at the inlet of a pump circulating or delivering the liquid.

Such an occurrence should be avoided by ensuring that adequate pressure exists at all times at pump inlet. Cavitation is generally accompanied by a high-pitched noise. It will reduce the pump delivery and lead to serious damage to the pump.

**Cell furnace:** this is a chamber constructed from refractory into which a fuel is fed and burned. (See Fig. 15.1.)

**Characteristic curve:** a graph relating one measurement of the performance of a machine to another. In this book the term is used to relate power required, pressure generated and efficiency to the rate of fluid flow through a pump or fan. (See Fig. 7.1, 7.4, 7.5 and 7.6.)

**Damper:** a device placed in a duct and used to control the air or gas flow in that duct by creating a variable resistance to flow. The variation of resistance is achieved by varying the gas flow area by altering the position of the damper.

**Dimensionless numbers:** groups of symbols which represent the properties of a system of physical change such as fluid flow, heat transfer and mass transfer, such that, when values are assigned to them, the final result is independent of the system of measurement used, provided that all items are in self-consistent units. (See also **Nusselt number, Prandtl number** and **Reynolds number**, and Section 5.4.)

**Downcomer:** a tube or pipe in a water tube boiler which conveys water from an upper reservoir or drum to a lower reservoir (drum or header) in the course of natural circulation promoted by heating the water in other tubes in the same hydraulic circuit. (See also **Riser** and Section 5.11.)

**Dry mineral matter free:** this defines a basis for the chemical analysis of a fuel. The proportion of constituents present is related to the total mass of the sample minus the sum of the amounts of water and all incombustible material present. It is also known as 'Parr's Basis'. (See Section 3.3.)

**Evasé:** a section of ducting in which the cross-sectional area of the fluid inlet is less than that of the fluid outlet. It is used to reduce the gas velocity in the duct and hence to convert kinetic energy into pressure energy. (See Section 7.2.4.)

**Firetube boiler:** a boiler in which the products of combustion or hot gas flow through ducts, mostly of tubular form, which are wholly contained within a water-cooled vessel. Combustion may also take place within a large duct also enclosed in that vessel. (See Chapter 1.)

**Flame:** the combination of combustible gases with a reactant, usually air, leading to the evolution of heat. This is accompanied by visible radiation and significant electrical conductivity.

**Flame envelope:** the outer regions of a flame beyond which there is no significant chemical reaction.

**Flame front:** that position of a flame which approaches and tends to move in the opposite direction to the incoming unignited mixture of atomised fuel or gas, and air.

**Flammability:**   the ability of a mixture of a substance and a reactant to form a flame when a suitable high-temperature source of energy is presented to it. The term 'flammability', however, has a specific meaning in combustion science. See *Technical Data on Fuel* (7th edn), 1977 Ch. 5 (British National Committee, World Energy Conference, London, SW1A 1HD).

**Heating value:**   See **Calorific value**.

**Ignition plane:**   the area over which the ignition of freely charged solid fuel takes place when such a fuel is burned on a grate. It is analogous to the flame front when gas or atomised liquid fuel is burned.

**Interstage spray (Desuperheater):**   a device used to reduce the temperature of steam, in which water is sprayed into the steam. The device is usually placed between two sections or stages of the superheater, e.g. between the primary and secondary sections (see Fig. 2.12). The water is evaporated by the transfer of heat from the relatively high temperature steam, thus cooling the steam and contributing to the steam flow through the secondary section.

**Iris damper:**   a damper is a device placed in a duct and used for controlling the flow of a gas in that duct. An iris damper is a particular type of that device in which the control is exercised, in a circular duct, by sectoral plates hinged so that they can move over one another, thereby increasing or decreasing the control area through which the gas flows. It is a similar mechanism to the iris of an eye or the aperture adjustment in a camera lens.

**Ligament:**   the minimum distance between adjacent holes made in a plate into which tubes are fixed (see Fig. 2.4).

**Links:**   the individual elements of a chain. A chain-grate stoker comprises a mattress of flat links joined together by rods which pass through holes in the ends of the links (see Fig. 14.11).

**MCR:**   this is an abbreviation for 'maximum continuous rating', which is the maximum output, in terms of heat units (or mass per unit time of steam) on a continuous basis, for which a boiler is designed.

**NPSH:**   this is an abbreviation for 'net positive suction head'. It is an important factor in the design of any pumping circuit. It defines the positive pressure available at the inlet to the pump after deduction of the pressure losses due to acceleration and friction in the inlet pipe system under conditions of maximum intended flow rate. The pump supplier will define the minimum NPSH at which the pump can operate without damage by cavitation. See also **Cavitation**.

**Nusselt number:** a dimensionless number related to the heat transfer properties of a fluid system. It is given by

$$Nu = h \, d \, k^{-1}$$

where: $Nu$ is the Nusselt number;
       $h$ is the heat-transfer coefficient;
       $d$ is a characteristic dimension of the system in which flow takes place; and
       $k$ is the thermal conductivity of the fluid.
(All in self-consistent units.)

**Orifice plate:** a plate provided with a central hole, which is fixed in a duct or pipe through which a fluid is flowing. The purpose of the device is to cause a pressure drop, either to reduce the flow rate to a predetermined value or to provide a means of measuring the flow by inference from a measurement of differential pressure in the pipe or duct.

**Oxygen scavenger:** a chemical reducing agent introduced into the boiler water to combine with and render innocuous any oxygen dissolved in the water. The substances generally used are sodium sulphite or hydrazine. (See Section 6.8.2.)

**Pile burning:** a process in which a material is supplied in such a manner that a conical heap is produced and combustion takes place on the surface of the cone.

**p.p.m.:** an abbreviation for 'parts per million'. (See Chapter 6.)

**Prandtl number:** a dimensionless number related to the properties of a substance. It is given by

$$Pr = C_p \, \mu \, k^{-1}$$

where: $Pr$ is the Prandtl number;
       $C_p$ is the specific heat at constant pressure of the substance;
       $\mu$ is the viscosity; and
       $k$ is the thermal conductivity.
(All in self-consistent units.)

**Pressure part:** that part of a boiler or pressurised system which has been designed to withstand a given pressure. In firetube and water-tube boilers this usually refers to the parts containing steam and/or water.

**Primary air:** that part of the air supply to a combustion system which the fuel first encounters.

**Pulverized fuel:** solid fuel which has been ground down to fine particles which are then burned in suspension using burners similar to those used for oil or gas. Analogous to **Atomisation**, which is a term used principally for liquid fuels. (See Section 14.2.2.)

**Quarl:**   a refractory structure which surrounds the discharge end of a burner for gas, oil or pulverised fuel. It serves to protect metal parts of the burner from excessive heat and at the same time provides a hot environment for the stabilisation of the flame front.

**Rapping gear:**   a mechanism for imparting regular mechanical shock to parts of a dust collecting system (generally an electrode of an electrostatic precipitator, or a bundle of tubes) whereby dust can be dislodged from the surface on which it has collected. The dust then falls into a hopper whence it is discharged as bulk powder to a removal system.

**Register:**   that part of a gas or oil burner in which turbulence is given to the air flow immediately prior to mixing with the fuel.

**Restrictor:**   a device for restricting the flow of fluid in a pipe. It may consist of an orifice plate or a hole in an elongated plug in a pipe. It is used in this book to describe a device for regulating the flow of water through tubes in a water-tube boiler.

**Reynolds number:**   a dimensionless number related to the state of flow of a fluid in a duct or pipe. It is given by

$$Re = D\,G\,\mu^{-1}$$

where:  $Re$   is the Reynolds number;
$\phantom{where:}$ $D$   is the characteristic dimension of the duct, e.g. its diameter;
$\phantom{where:}$ $G$   is the mass flow rate of the fluid; and
$\phantom{where:}$ $\mu$   is its dynamic viscosity.
(All in self-consistent units.)

**Rifled bore:**   the bore of a tube in which there are helical grooves to increase the turbulence of a fluid flowing through it.

**Ringelmann number:**   a measure of smoke density depending on visual observation of the smoke in comparison with four cards where black lines at right angles to each other obscure a white background in standard percentages of obscuration. (See BS 2742.)

**Riser:**   a tube in a water-tube boiler in which the water, or steam and water mixture is heated, thereby losing density. As a consequence, dense water in unheated tubes in the same hydraulic circuit causes the heated mixture to rise. The term also refers to the tubes or pipes connecting boiler upper water-wall headers to the drum to convey the steam produced to the drum for separation from the water.

**Secondary air:**   the second stage of admission of air to a combustion system, generally to complete combustion initiated by the primary air. It can be injected into the furnace of a boiler under relatively high pressure when firing solid fuels in order to create turbulence above the burning fuel to ensure good mixing with the gases produced in the combustion process and thereby complete combustion.

**Singles coal:**   coal particles sized between 13 mm and 25 mm.

**Smalls coal:**   graded coal with no lower size limit, but with an upper size of 50 mm.

**Stay bar:**   a solid bar of metal connected to the opposite sides of a pressurised enclosure to give mutual support to those sides, e.g. the two tube plates of a firetube boiler.

**Stay tube:**   a metal tube, generally conveying hot gases, connected to the opposite sides of a pressurised enclosure to give support to those sides. Such a tube has a greater wall thickness than those tubes whose purpose is solely to convey gases.

**Stoichiometric:**   'that which is just sufficient for'. In chemistry this refers to that quantity, and no more, of a substance theoretically required to complete a reaction with another. In combustion technology, stoichiometric air is that quantity of air, and no more, which is theoretically needed to burn completely a unit quantity of fuel. 'Sub-stoichiometric' refers to the partial combustion of fuel in a deficiency of air.

**Stoker:**   a term previously used to denote a person who attends to the burning of solid fuel in a furnace or boiler. In this book it is used as an abbreviation for 'automatic stoker', which consists of a machine for conveying solid fuel to and firing it in a boiler.

**Superheater:**   a tubular heat exchanger placed in the path of travel of hot gases in a boiler and through which steam passes. The purpose of a superheater is to raise the temperature of the steam generated in that boiler.

**TDS:**   an abbreviation for 'total dissolved solids'.

**Tertiary air:**   a third stage of admission of air to a combustion system, the reactions of which have largely been completed by secondary air. Tertiary air is rarely needed.

**Toroidal:**   a form of eddy in a fluid stream in which the axis of the eddy is parallel to the circumference of the duct in which the fluid flows.

**Traversing chute:**   a near-vertical duct through which coal is led from a storage bunker at the front of a boiler to a travelling grate stoker on a water-tube boiler. The lower end of the chute continually moves slowly from one side of the boiler to the other, thus traversing and distributing coal over the full width of the stoker (see Fig. 2.3.).

**Tube seat:**   the joint between a hole in a metal plate and the tube fitting and sealed into it (see Fig. 2.4.).

**Turndown:**   the extent to which the throughput of a fuel-firing appliance can be reduced, without difficulty, below the maximum throughput for which it has been designed.

**Turndown ratio:**  turndown expressed as a ratio; a turn down of 4 : 1 means that the throughput of the device can be reduced to a quarter of the maximum

i.e. $\dfrac{\text{maximum load}}{\text{load at turn down}} = 4$

**Velocity head:**  the energy in a fluid-flow system due to velocity. It is given by

$$h = \frac{V^2}{2g}$$

where  $h$  is the velocity head in units of height of the fluid under consideration;

$v$  is the velocity of the fluid; and

$g$  is the local acceleration due to gravity.

(All in self-consistent units.)

**Venturi:**  an elongated restriction placed in a duct to create a pressure differential which can be measured and translated to give a measure of flow through the duct.

A venturi is formed by reducing the cross-sectional area of the duct with tapered sections at the inlet and outlet, the outlet taper being more gradual than that of the inlet. A venturi designed to BS 1042 gives a large recovery of the measured pressure differential such that the net energy loss is low compared with that across an orifice plate of equal measured differential.

**Volute:**  a fan casing where the distance from the inside surface of the casing to the fan axis increases continuously from a minimum to a maximum in the direction of the rotation of the fan blades. The maximum distance occurs at the fan outlet.

**Water-tube boiler:**  a boiler in which the water is contained in and circulates in an arrangement of tubes of relatively small diameter.

**w.g.:**  an abbreviation for 'water gauge', a measure of pressure as head of water which is used for air and gas within the gas passages of a boiler. A water gauge is also used to indicate the level of liquid in a vessel, and, in particular, as a mandatory requirement to indicate the level of water in a steam boiler drum.

**Windbox:**  that part of a firing appliance for a boiler into which the air for combustion is received and from which it is directed into one or more firing appliances.

# CHAPTER 1

# Boilers, their evolution, types and applications

## 1.0 Early boilers

Boilers, be they in power stations, breweries, schools, or dwellings, are accepted equipment of everyday life. Their history, however, is a long one: Hero of Alexandria invented a combined reaction turbine and boiler about 100 BC, and domestic boilers were known in Pompeii in the first century AD.[1,2]

Just as the purpose of Hero's *Pneumatica* was to provide mechanical power so, at the beginning of the industrialisation of Britain in the early eighteenth century, a need was felt to supplement human effort by mechanical power. The Marquis of Worcester, Savery, Newcomen, and Watt are all names associated with the development of steam power, and therefore with boilers.

## 1.1 Firetube boilers

Early boilers consisted of closed vessels made from sheets of wrought iron which were lapped, riveted and formed into shapes varying from simple spheres to complex sections such as the 'waggon' boiler of Watt (1788) illustrated by Fig. 1.1, so called because its shape resembled that of a covered waggon. These vessels were supported by brickwork over a fire which itself was supported on a grate. To make use of the surface not seen by the fire the products of combustion were circulated over much of the surface by means of flues formed in the brickwork, as illustrated. Such boilers were said to be 'externally fired'. A serious disadvantage with this system of firing was that scale and sludge were precipitated from the water to the bottom of the boiler immediately over the fire and the hottest gases. This material insulated the metal from contact with the water which would

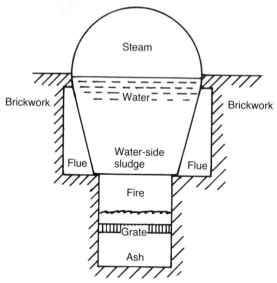

**1.1** Watt's 'waggon' boiler

otherwise have kept it at a safe temperature. In the absence of a deposit of this kind the metal in the bottom of the boiler would have been at about 200 °C for a working pressure of 0.7 bar (10 lb in$^{-2}$) which was typical in those days. With 5 mm of scale and sludge the temperature of the metal would have been over 500 °C (see Fig. 6.7). The metal would have then distorted and, even at this low pressure, ruptured disastrously.

The demand for more powerful engines created a need for boilers which would operate at higher pressures. This required new thinking to avoid the problem just discussed. Trevithick in Britain and Evans in the USA each conceived the idea of internal firing. This was done by making the pressure vessel cylindrical and providing an internal cylindrical furnace submerged beneath the water and large enough to accommodate a grate on which to burn the coal. The products of combustion passed from the fire along the flue, turned downwards to pass beneath the boiler to the front end, then the gases passed rearwards along the sides of the boiler through two brick flues, one either side, to a chimney. There were thus three 'gas passes' – the furnace, the bottom flue, and the side flues as illustrated by Fig. 1.2. Loose sludge would indeed still fall to the bottom of the boiler, but this was now relatively gently heated, not by live fire, but by products of combustion which had lost much of their heat content in the furnace. There was thus less risk to the bottom of the boiler, although scale could form on the furnace and endanger that. This was to remain a problem for many years and was not eradicated until the chemistry of water treatment was understood (see Ch. 6). In spite of this, Trevithick's 'Cornish' boiler could operate at pressures up to 7 bar, i.e. ten times greater than the

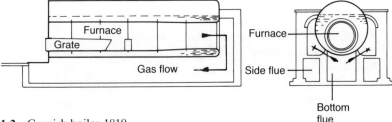

**1.2**   Cornish boiler 1810

earlier boilers. These internally fired boilers became known as 'shell' or 'firetube' boilers.

It was a short step from the single-furnace Cornish boiler to the larger, twin-furnace 'Lancashire' boiler patented by Fairbairn and Hetherington in 1844. This is illustrated by Fig. 1.3 and dominated the industrial steam-raising scene until the early 1950s. Possibly more than a thousand of these boilers are still working in the UK today[3] although they are being replaced by the more efficient multitubular types known as 'Economics'.

It will be appreciated that the greater the area of boiler surface exposed to heat transfer, the greater will be the amount of heat available from a given consumption of fuel, i.e. the efficiency of heat recovery will be greater. A large number of small-diameter tubes conveying hot gases through the body of water in the boiler provides such an increase in heated surface, and at the same time avoids the need for flues in the foundation and at the sides of the boiler. Such a boiler becomes 'self-contained'. As will be seen in Section 12.4.1 the longer the tubes and the smaller their diameter the greater is the effectiveness of the heat transfer surface. These multi-tubular boilers are therefore much more compact for a given output than were their predecessors, and do not require to be set in brickwork. They found extensive application on ships and for locomotives where, in both cases, space is at a premium.

Some early marine boilers were rectangular in section in order to make more effective use of the space available. Serious explosions occurred due

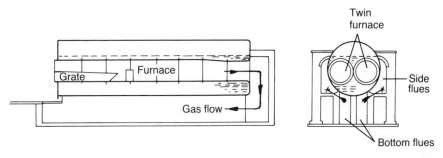

**1.3**   Lancashire boiler 1844

to rupture at the corners. A vessel subject to internal pressure will tend towards a spherical shape, and other shapes will therefore be highly stressed in those areas which depart from the spherical. The nearest practical approach to the sphere, for boilers, is the cylinder, especially if the ends are domed. A design for a spherical boiler was actually patented in the early 1800s with the intention of being able to withstand pressures up to 200 bar. It was specified to be made from copper, 64 mm thick, but there is no record of it ever having been made. It was to have been externally fired, which in itself presents problems, but in the geometry of a sphere it would not seem to be feasible to use internal firing. Even with present-day alloy steels it is not permitted anywhere in the world in firetube boilers to expose metal more than 22 mm thick to fire or high-temperature gases. This is to avoid excessive thermal stress in the metal.

Marine boilers were eventually all made cylindrical, but brickwork settings and external flues were quite unacceptable on board ship due to their size and weight. Multi-tubular boilers were therefore used and internal furnaces, up to four in number, were retained. The gases from the furnaces entered individual water-walled reversal chambers then, turning through 180°, entered a multiplicity of tubes each of about 75 mm bore. After passing through these, the gases entered the funnel. These were 'two-pass' boilers. Later, a three-pass version evolved in which the gases passed from the front to the back of the boiler through a further set of tubes. These boilers were known as 'Scotch Marine Boilers' and persisted from around 1850 until diesel power replaced steam.

The Scotch boiler was slowly adapted for land use, but as there was less restriction on space it was made longer. In the UK, the generic name is 'Economic' but in the USA it is still known as the 'Scotch' boiler. Being smaller, cheaper and inherently more efficient than the Lancashire type, it began to compete with the latter in the early 1930s. In the land-based boilers, the reversal chamber was refractory lined but in a later development the water-walled chamber was revived. Figures 1.4(a) and 1.4(b) illustrate, respectively, two-pass and three-pass Economic boilers with refractory reversal chambers, which are termed 'dry-back' boilers.

The three-pass Economic boiler suffered from a problem which arose from the use of a single end plate into which the tubes are connected (the tube plate) to accommodate both second- and third-pass tubes. The gases entered the second pass at about 1000 °C and left the third pass at about 250 °C. The tube plate was therefore subject to two different temperature conditions, which set up stresses and which led to leakage at the tube seats. In 1935 the Lincoln firm of Ruston and Hornsby patented[4] a three-pass construction based on the three-pass Scotch marine boiler. This overcame the problem of a single tube plate being subjected to two different temperature regimes by providing separate tube plates for each condition (see Fig. 1.5). The water-walled reversal chamber was used, the front wall of which accommodated the furnace exit and the entries to the second-pass

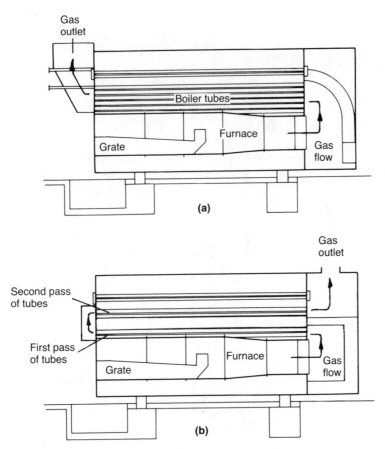

**1.4**  (a) Two pass and (b) Three pass dry-back Economic boilers

tubes. This water-walled, or 'wet-back', construction had the added advantage of replacing the heat-losing refractory linings of the earlier boilers with effective heat receiving surfaces. After passing to the front of the boiler, the gases reversed in the 'front smoke box' and returned to the rear of the boiler through tubes forming a third pass, these tubes being seated in an independent tube plate at the rear. The front tube plate of the boiler accommodated the outlets of the second-pass tubes and the inlets of the third pass, but there was no tube plate differential temperature problem since the gases left the second pass and entered the third pass at virtually the same temperature. This construction is now widely used throughout the world.

The next major development took place in the USA. During the Second World War a need arose for boilers to supply steam to field installations. It was essential that their installation and commissioning should take place in a minimum of time. To this end the boiler was supplied, not as hitherto as

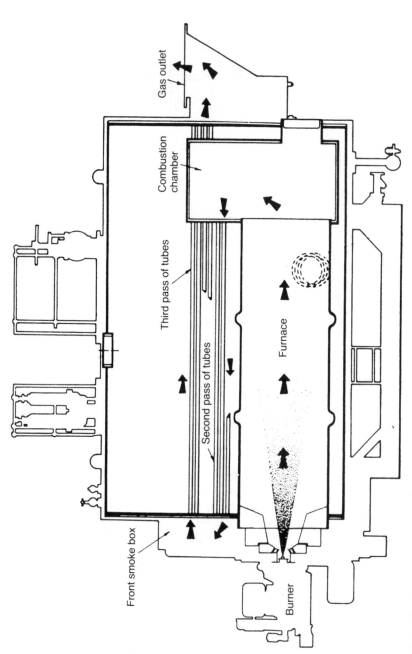

Gas outlet

Combustion chamber

Third pass of tubes

Second pass of tubes

Furnace

Front smoke box

Burner

**1.5** Wet-back three-pass Economic boiler

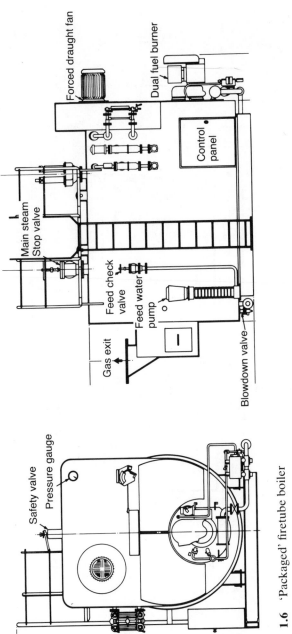

**1.6** 'Packaged' firetube boiler

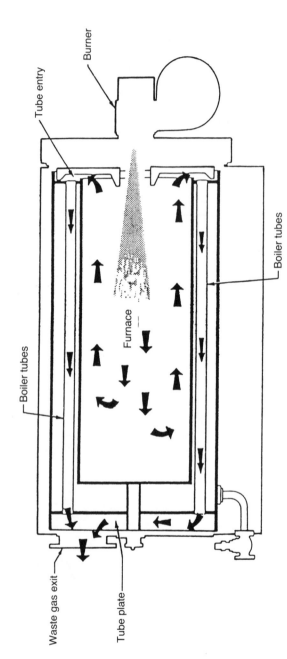

**1.7** Reverse-flame boiler

a shell only to be fitted with firing equipment, pumps, valves and fans on
site and from separate suppliers, but as a working entity complete with all
auxiliaries. This was the so-called 'Package' boiler and consisted of a three-
pass, dry-back, Scotch boiler mounted on a base frame. After the war this
Package idea became very popular and all but the very large firetube
boilers made today are of this concept. The boilers are delivered complete
with all auxiliaries, as illustrated by Fig. 1.6; they are factory assembled,
tested and ready for work, and are mainly of the three-pass wet-back type.

For smaller boilers, the reverse-flame type of construction is often used,
particularly for water heating. In this type the furnace has only one open
end, the burner is positioned on the axis of the furnace firing towards the
closed end, and the gases return concentrically round the outside of the
flame as illustrated in Fig. 1.7. The single pass of boiler tubes is arranged
concentrically around the furnace. Since there is only a single pass of
tubes, turbulence promoters are fitted to increase the heat transfer and to
reduce the exit gas temperature.

The rear end of the furnace is inactive so far as gas flow is concerned
since, due to the blank end, pressure builds up. It is important that the
burner is designed to give a long, narrow, penetrating flame; a short, bushy
flame will tend to be picked up by the outgoing gases and this will lead to
burning gases entering the tubes, thus increasing the metal temperatures in
this region.

Figure 1.8 compares the sizes of a Lancashire boiler and a modern boiler

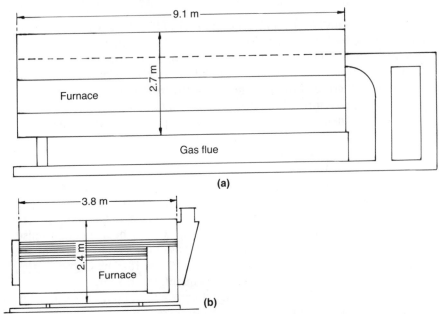

**1.8** Comparison in size between (a) Lancashire and (b) modern boilers for the
same output of 3.4 MW. From the Institute of Energy

for the same output of 3.4 MW. The former would have had an efficiency of around 65% based on the gross calorific value of the fuel when combustion conditions were good, the latter an efficiency of about 80% under the same conditions, the difference representing a fuel saving of approximately 20%.

## 1.2   Water-tube boilers

As industry developed during the last century, so the use of boilers for raising steam became widespread and, for various reasons discussed in Chapter 9, disastrous explosions sometimes occurred. Boilers of that period consisted of heated pressure vessels of large diameter and subject to internal pressure which set up tensile stresses in the walls of the enclosure. The value of this stress, known as 'hoop stress' is given by

$$f = \frac{P \times D}{2T}$$

Where:   $f$ is the hoop stress;
$P$ is the internal working pressure;
$D$ is the vessel diameter; and
$T$ is the thickness of the metal.
(All in self-consistent units).

It will be seen that, for a given stress $f$, as $D$ increases with increasing output of the boiler, $T$ must also increase. If the working pressure $P$ increases, then either $D$ must be decreased or $T$ be increased to keep $f$ within acceptable limits. If $T$ is increased then the mass of the boiler and its cost of manufacture both increase. The attractive alternative is to decrease $D$. This approach formed the basis for the water-tube boiler in which the water is contained within tubes and the gases pass across the outside of them.

There were a number of early designs of water-tube boiler developed throughout the late eighteenth and the nineteenth centuries.[2,5] As is frequently the case with development today, the rate of progress seemed to have been restricted by the availability of tubes and suitable materials to withstand the higher operating pressures that were being aimed at.

The forerunners of today's designs appear to have been a compromise between a firetube and a water-tube boiler as developed by Stephen Wilcox in 1856,[5] a straight-tube boiler patented in 1867 by George Babcock and Stephen Wilcox, and subsequently in 1877 by a design of the type illustrated by Fig. 1.9. With the latter design of boiler the water-filled tubes were inclined and the ends belled or 'expanded' into headers at each end, which were in turn connected to a cylindrical vessel called the steam drum. This drum contained water having a space above, into which the steam generated in the tubes was released before being discharged to the steam consumer.   *why*

The boiler-heated surfaces consisted of a group or 'bank' of tubes of about 75 mm bore, some of which were exposed to the fire, the others to the flow of hot gases produced by the combustion process. Baffles were provided in the bank of tubes to create a number of gas paths and thus to increase the effectiveness of the heated surface, as illustrated by Fig. 1.9. In this way the heat was transferred to the water in the boiler through tubes of relatively thin section when compared with the thickness of a firetube boiler shell. This being so, the working pressure could be raised considerably above that possible in a firetube boiler. Moreover, should a

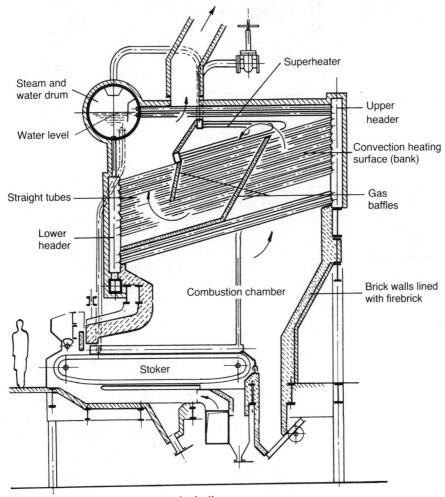

**1.9**   Early straight-tube water-tube boiler

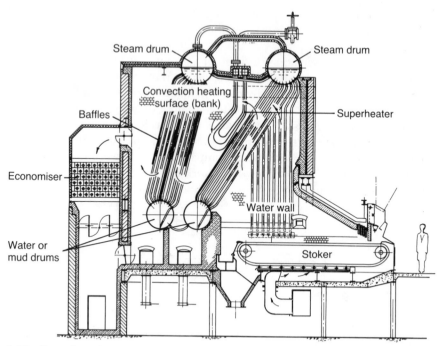

**1.10** Four-drum bent-tube water-tube boiler with partially cooled walls

tube rupture occur, the consequences would be less serious than if the furnace or shell of a firetube boiler ruptured.

While this construction persisted for many years, it had pressure limitations, and eventually multidrum boilers with bent tubes were developed, one such design being illustrated by Fig. 1.10. The lower drums into which boiler water impurities settled became known as 'mud drums'.

In the early boilers, the tubes were situated in an enclosure made of brickwork lined on the inside with firebrick, as illustrated by Fig. 1.9. Eventually, the brick walls were partially covered on the fire side by water-tubes, called water-walls, connected to the steam drum direct or by headers and pipes as illustrated by Fig. 1.10. These absorbed heat from the combustion process, thereby reducing the temperature of the gases flowing to the tube bank (section 5.2), gave some protection to the brick walls and eventually, as the percentage of the wall area covered by tubes increased, gave scope for reducing the heat loss from the walls to atmosphere.

Designs progressed until, at present, the walls are usually completely covered with water-cooled surface (Fig. 1.11). These walls are formed either by tubes touching each other, called 'tangent tubes', or by tubes connected together by strips of steel welded axially between them called 'welded walls' (Fig. 5.1(c)).

With increasing working pressures and the use of economisers and

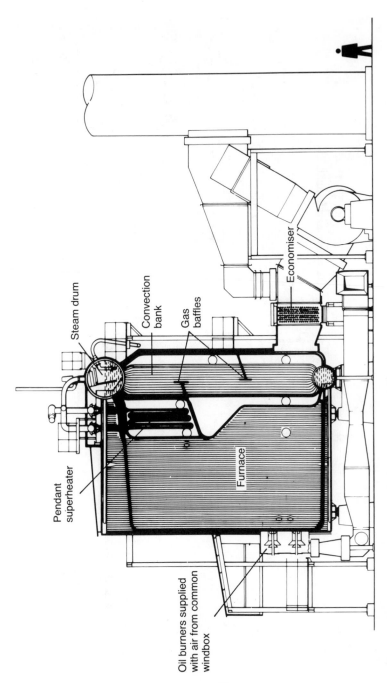

Steam drum

Convection
bank

Gas
baffles

Economiser

Pendant
superheater

Furnace

Oil burners supplied
with air from common
windbox

**1.11**   Two-drum oil- or gas-fired boiler with fully cooled walls

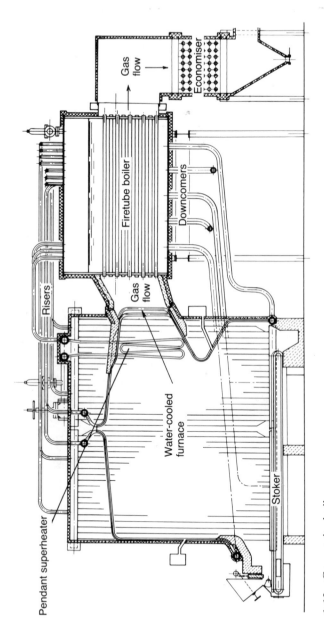

**1.12** Composite boiler

airheaters (Section 7.4) it has become more economical to reduce the number of boiler drums, the two-drum type of boiler now being the most popular design among industrial water-tube boilers operating at pressures up to 100 bar, although a number of single-drum types are available (Section 2.3)

Recently, the 'Composite' boiler has been developed in which the water-tube and firetube principles are combined (Fig. 1.12). With this design, the steam drum and convection bank of a conventional water-tube boiler are replaced by a shell containing a large number of small-bore tubes only, i.e. a firetube boiler with no furnaces. The water-walls of the furnace are connected by pipes to the firetube boiler which acts also as the steam drum. The design combines the advantages of the low-cost firetube boiler with its large steam/water drum and consequent low steam disengagement velocity, with the ability to arrange the water-tube furnace with its flexible combustion space so as to incorporate the type of firing equipment most suitable for a particular fuel and application.

## 1.3  Applications

The purchaser of boiler plant has the option of choosing the firetube, the water-tube, or the composite types. Each of these has quite distinct characteristics, and although firetube boilers are inherently cheaper than water-tube boilers for the same output and pressure, it is the purpose for which the energy output is required that will determine which type is used.

It has been estimated[3] that in 1979, there were over 73 000 steam boilers (excluding utility boilers) operating in the UK, of which 61% were horizontal firetube boilers, 5% water-tube, the remainder being 'vertical' and 'brick flue' (the last becoming rapidly obsolescent). To these must be added a considerable but unstated number of hot-water boilers operating at medium- and high-water temperatures. These are used mainly for space heating, but also for some process applications. They are likely to be mainly of the firetube type, although water-tube boilers of the 'La Mont' forced-circulation type (see Section 2.3.3) are used. It looks as though there could be about 100 000 industrial boilers in operation in the UK, 60 000 firetube and 5000 water-tube, the rest being obsolescent.

The most common type by number is obviously the firetube boiler, but by capacity water-tube boilers account for 29% of the total[3], which suggests much larger unit outputs for this category. Indeed, it is known that most water-tube boilers are concentrated in those large industrial complexes operating continuously throughout the year where demands for power and heat are in good balance, e.g. chemical works, oil refineries and the steel industry. For power production, high pressure and temperature are needed, often beyond the scope of firetube boilers. Where heat only is needed, e.g. space heating, in breweries, in laundries and in the food industries, indeed in most of the general run of industries, the firetube boilers suffice – this accounts for their preponderance by number.

## References

1. Murray, M. V. 'The shell boiler, an historical review', *J. Inst. Fuel* 1959 (Sept.) p. 425.
2. Thompson, Colonel S. J. *Boilers – Past and Present*. Inst. Mech. E., 1942.
3. Robson, M. and Chesshire, J. *Energy Use For Steam Raising in UK Industry and Commerce*. Science Policy Research Unit, University of Sussex, 1982.
4. British Patent No. 451, 770.
5. *Steam, its Generation and Use*. Babcock & Wilcox, New York, USA, 1972.

# CHAPTER 2

# Boiler options

## 2.0   Selection of boiler type

The selection of the type of boiler to be used for a particular application depends among other things upon the prevailing limitations in the capacity and steam pressure and the temperature capability of the various designs available. These can vary from one manufacturer to another and will depend upon the types of boiler in which the manufacturers specialise, hence any limiting steam conditions given in this chapter can only be typical for the Industry in general (Table 2.1.) It may, for example, be found that one manufacturer will offer a firetube boiler where others will offer a small water-tube one because they do not manufacture the firetube type.

Costs can also vary from one supplier to another, making definite statements regarding the selection of boilers impossible. The comparisons in this chapter are made on the basis of all types of boiler being available from one source.

Perhaps the major factor in plant selection is initial capital cost. When a number of tenders are received at an equal price and extent of supply, the engineering features must then be evaluated to obtain the optimum design. Factors for consideration are – auxiliary power consumption, ease of cleaning, the heat transfer rates in the various parts of the boiler, the quality of feedwater required (Ch. 6) and the capabilities of the operators available.

The types of boiler available for industrial use are:

- firetube or shell;
- composite (firetube and water-tube combination); and
- water-tube.

**2.1** Characteristics of firetube and water-tube boilers

|  | Firetube | Water-tube |
|---|---|---|
| Pressure | As a fired boiler it is limited to 20–30 bar (20 in the larger sizes). As a waste heat boiler it can be considerably higher. | Virtually unlimited. |
| Unit output | Limited to about 20 MW. | Virtually unlimited. |
| Fuel acceptance | All commercial fuels and certain treated wastes. | Virtually unlimited due to the large furnace which can be designed for a particular fuel. |
| Cost | Low compared with water-tube in areas where duties overlap (see Fig. 2.13). | High compared with firetube in areas where duties overlap (see Fig. 2.13). |
| Erection | Packaged ready for work after connecting to services at site. | Can be shop-assembled or site erected. |
| Efficiency | 80–85% (gross calorific value) according to fuel. Can be increased by adding economiser. | 85–90% (gross calorific value) according to fuel. Economiser or airheater normally included as standard. Both may be used together to maximise efficiency. |
| Water | Necessary to BS2486. | Necessary to BS2486 but more exacting than for firetube. |
| Main purpose | To supply heat. | To supply power plus heat. |
| Inspection | Normally every 14 months. | Normally every 26 months. |

The last can be subdivided by method of construction into:

- shop-assembled;
- modular; and
- site-assembled.

## 2.1 Firetube boilers

These may be vertical or horizontal, the majority being horizontal, although recent developments in fluidised bed combustion favour the vertical arrangement due to the generous space or freeboard above the fuel bed (see section 14.2.3) provided in the furnace. One factor limiting the output of vertical boilers to about 3.5 MW is the restricted steam release space available. Another difficulty is the accommodation of the heated surfaces in a vertical shell, which can prove to be very expensive.

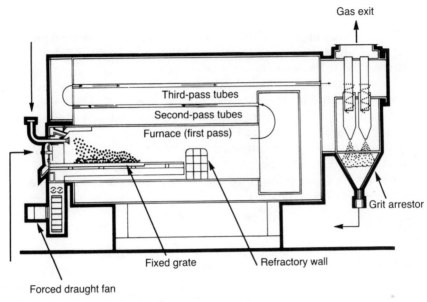

Gas exit

Third-pass tubes

Second-pass tubes

Furnace (first pass)

Grit arrestor

Fixed grate

Refractory wall

Forced draught fan

**2.1**   Firetube boiler with fixed grate

A typical horizontal firetube boiler is illustrated by Fig. 2.1. This consists of a horizontal furnace, which may vary from 0.5 m to 1.8 m in diameter according to the output required from the boiler. For outputs beyond about 10.0 MW (oil- or gas-firing) or 6.0 MW (coal-firing), two furnaces placed side by side are used. Firetube boilers are sold as pre-engineered standard ranges, but specials can be made to suit unusual circumstances.

Because the whole of the heat-transfer surfaces of a firetube boiler are enclosed in a shell, which is also the steam and water drum, this is necessarily large in relation to the boiler output. It provides generous steam space and water capacity, both of which are advantageous for meeting heavily peaked loads. Indeed, the shell size may be increased beyond that necessary to accommodate the heated surfaces to enhance this effect; this is the basis of the 'thermal-storage boiler' (Section 16.8.3.).

### 2.1.1   Limitations of firetube boilers

#### 2.1.1.1   *Size and output*

Any load transported by road in the United Kingdom is normally limited to 4.3 m wide, although this is subject to conditions on the particular route chosen. This limits the output of current designs of firetube boilers to about 20 MW (oil- or gas-fired) and 12 MW (coal-fired) per boiler. For special contracts it may be possible to obtain a relaxation of these restrictions from the Department of Transport.

### 2.1.1.2 Pressure

The cylindrical furnace of a firetube boiler is subject to external pressure which can lead to collapse. Increasing the thickness increases thermal stresses due to the temperature gradient across the material, and most national standards now limit the maximum thickness to 22 mm and provide formulae by which the safe working pressure can be determined. In practice, this may vary from about 30 bar for small-diameter furnaces to 18 bar for large furnaces, see Fig. 2.2. The furnace is therefore the main limitation to the pressure for which a firetube boiler can be built. Waste-heat boilers (Ch. 16), where no furnace is required, can be built for considerably higher pressures. This also applies to the composite boiler illustrated by Fig. 1.12

### 2.1.1.3 Fuel

The furnace of a firetube boiler, being limited in size, is limited in its ability to accommodate the firing appliance, particularly for certain bulk fuels, for example palm fruit waste and bagasse, the waste from cane sugar factories. It can, however, accommodate suitable firing appliances for normal commercial fossil fuels.

It emerges, therefore, that for unit outputs exceeding the capability indicated by Fig. 2.2 the option is clearly for composite or water-tube type boilers. For all other purposes the cheaper and more easily installed and managed firetube boilers will suffice.

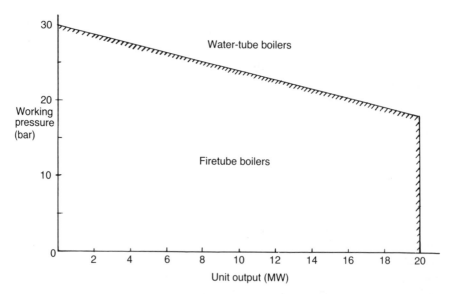

**2.2** General scope for basic boiler types

As it is now usual for firetube boilers to be delivered to site as a packaged unit, they can theoretically be delivered, emplaced, connected to services and set to work within days. This can and does happen, but in many cases extraneous matters, for example external labour and civil works, can delay this quite seriously. Coordination between contractors is therefore very necessary.

Firetube boilers can be equipped with superheaters and/or economisers. Superheaters are fitted when it is necessary to supply dry steam to consumers at the end of long pipe runs, or for those cases where modest electric power is to be generated. Economisers are now being widely used with gas-fired boilers, but problems can arise with oil firing and coal firing due to acid condensation (Ch. 10). Such problems can be controlled by careful design (Ch. 7) or by the use of neutralising additives in the fuel or combustion air supply (see section 10.4.2.2). When so fitted the efficiency obtainable from firetube boilers is comparable with that from water-tube boilers likewise fitted: 85% or more, based on the gross calorific value of the fuel.

## 2.2   Composite boilers, Fig. 1.12

The firetube boiler component does not include large-diameter flues, and this eliminates the major restriction in the pressure and output capabilities of firetube boilers. The steam output at which a composite boiler can operate is controlled by the manufacturing capability of the manufacturer of the shell boiler section and by transport restrictions. The maximum pressure will be limited by the tube-plate temperature, which is restricted by the design code BS 2790.[1] For solid-fuel firing the limitations are about 26 MW (11 kg s$^{-1}$) at 32 bar and 400 °C.

The composite boiler will usually be transported to site in sections, or modules, the major ones being the preassembled furnace and firetube boiler components.

## 2.3   Water-tube boilers

These can be of the single- or multiple-drum type. The design flexibility available with water-tube boilers has resulted in numerous configurations of drums and heating surfaces being used throughout the boiler industry worldwide. It is only possible in this book to refer to some of the most common designs now manufactured.

The majority of manufacturers currently favour either the two-drum or single-drum type.

## 2.3.1 Multi-drum water-tube boilers

Two or more drums are used to accommodate a bundle or bank of relatively closely spaced tubes giving a section of evaporative convection heating surface (Section 5.4.2) within the boiler to recover some of the heat from the hot gases. A typical four-drum type is illustrated by Fig. 1.10, and Fig. 2.3 illustrates a typical two-drum boiler, both arranged for coal firing.

   Expanded tube-to-drum connections are normally used in their construction, see Fig. 2.4(*a*). An expanded tube joint is formed by passing a tube through a hole in the drum only very slightly larger than the outside diameter of the tube and then increasing the diameter of the tube by using internal rollers or expanders which stretch and deform it, forcing it into

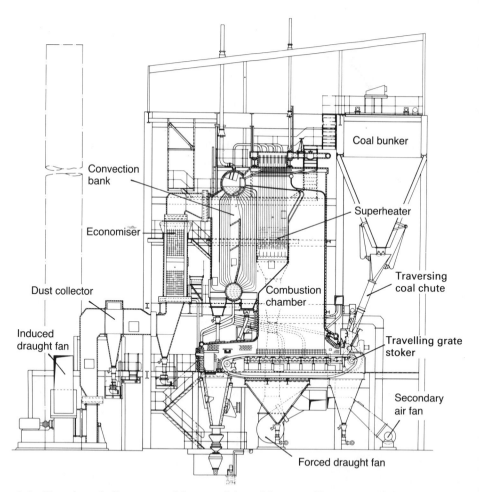

**2.3**   Two-drum boiler arranged for coal firing with a travelling grate stoker

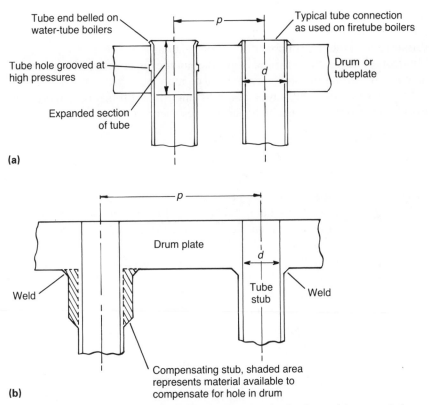

**2.4** Methods of attaching boiler tubes to drums and tubeplates: (a) expanded tubes; and (b) stub tubes welded to drum

tight metal-to-metal contact with the walls of the hole. To ensure that the tube stays in position it is expanded until the tube wall thickness is reduced by 5–10%. Boilers using this type of joint are restricted to working pressures below about 100 bar due to the difficulty of maintaining watertight joints at higher pressures.

At the higher pressures for which expanded tubes are used the holes may have circumferential grooves into which the tubes are deformed when expanded to increase the resistance to pull-out under pressure. The ends of the tubes are belled or tapered on completion in accordance with the boiler design codes,[2] see Fig. 2.4(a).

Evaporative convection heated surfaces in the form of banks of tubes become uneconomical at high pressures when high degrees of superheat are usually required. Because a high proportion of heat is transferred from the gases to the superheater, the gas temperature entering the convection bank is relatively low, and the convection bank is operating with a higher saturation temperature of the boiler water. Together these give a much reduced heat transfer compared with that with low-pressure boilers

(Section 5.6). This can be compounded by the fact that at high pressures it may be necessary to increase the spacing (pitch) of the tubes along the drums to maintain reasonable drum-plate thicknesses. This will have the effect of lowering the gas velocity andthe number of tubes, and hence the amount of heating surface that can be accommodated (Ch. 5).

For the drum of a water-tube boiler having rows of tube holes in it the expression given in Chapter 1 for hoop stress becomes

$$f = \frac{P \times D}{2T \dfrac{p - d}{p}}$$

Where: $P$    is the boiler design pressure;
       $D$    is the drum diameter;
       $p$    is the tube pitch – centre-to-centre distance – along the drum, see Fig. 2.4(a);
       $d$    is the diameter of the holes in the drum;
      $\dfrac{p - d}{p}$    is called the ligament efficiency and represents the proportion of drum material available to withstand the stress through a row of tube holes along the drum.

It can be seen from this that as the pressure $P$ increases, the drum thickness must increase if the stress $f$ is not to be exceeded, unless the value of $p - d$ is increased. The ligament efficiency can be increased by increasing the tube pitch or by reducing the tube hole diameter $d$.

For very high-pressure boilers the tubes are connected to the drums using extra thick stub tubes or connections welded to the drum as illustrated on Fig. 2.4(b). These compensate for some or all of the material removed by the tube hole, and efficiencies of 100% can therefore be achieved providing the requirements of the design codes for compensation are complied with.

### 2.3.2  Single-drum boilers

These can be used for any size of boiler and type of fuel and firing system. Single-drum boilers of natural, forced, or assisted boiler water circulation (Section 5.11) are used where there is a requirement for all welded tube joints, this feature being difficult to achieve with a two-drum boiler having closely spaced tubes in the convection bank, due to the limited access available between tubes for welding. Single-drum boilers are used almost exclusively for boilers operating at pressures above 100 bar and large boilers of single-unit outputs beyond the capability of the multi-drum type. A typical natural circulation design for industrial use having a high output at high pressure is illustrated by Fig. 2.5.

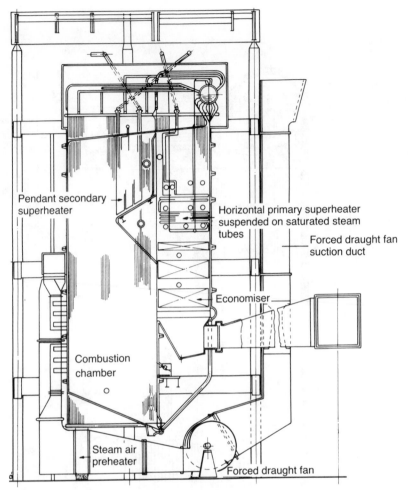

Pendant secondary superheater

Horizontal primary superheater suspended on saturated steam tubes

Forced draught fan suction duct

Economiser

Combustion chamber

Steam air preheater

Forced draught fan

**2.5**   Top-supported single-drum water-tube boiler

### 2.3.3   Forced- or assisted-circulation boilers

These are of the single-drum type using a pump to generate a major part of the differential pressure necessary to create the circulation of water through the boiler tubes, see Fig. 2.6. They are normally only used for pressures above 150 bar or at low pressures to special designs where the desired configuration of the heating surfaces does not lend itself to natural circulation, for example process waste-heat boilers (see Ch. 16).

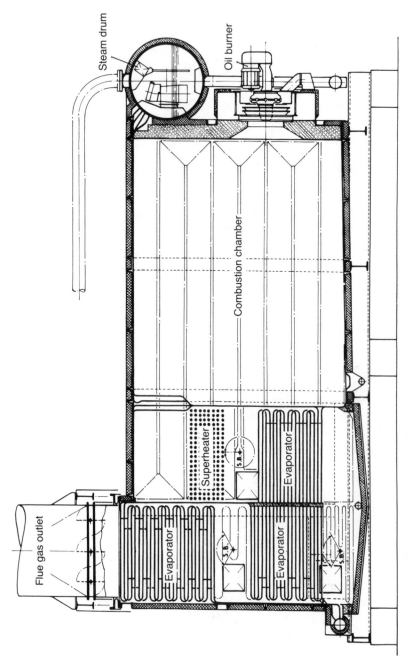

**2.6** Forced-circulation shop-assembled water-tube boiler

### 2.3.4  'Once-through' boilers

These are boilers where the water is pumped by the feedwater pumps
through a small number of tubes which form the economiser and
evaporator heated surface and, subsequently, the superheater (where one
is fitted). All but a small amount of the water is evaporated; the remaining

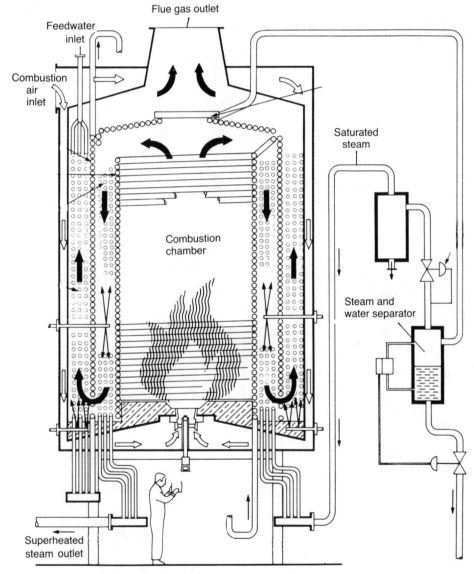

**2.7**  'Once-through type' coil boiler.
(*From M.E. Boilers Limited*)

water is separated from the steam in a separator vessel at the evaporator outlet and is discharged or blown down to dispose of the concentrated solids in the water (see Section 6.3). If required, the steam is then passed to the superheater.

There are a number of designs available for industrial purposes, one being illustrated by Fig. 2.7. Virtues of this type of unit are rapid start-up, since there are no large, thick pressure parts (drums) necessitating control of the rate of temperature raising, and the ability to operate with feedwater having a high total dissolved solids content (which simplifies water treatment).[3]

## 2.4 Methods of construction

Water-tube boilers can be transportable, modular or site-assembled.

### 2.4.1 Transportable, shop- or works-assembled boilers

The concept of shop assembly is that construction is under closely controlled conditions where all manufacturing facilities are readily available. The on-site requirement for specialists such as pressure-part welders may be reduced or even eliminated and the minimum on-site labour is required in situations where working conditions and facilities may leave something to be desired. For example, if the plant is outdoors (exposed to the weather, as in oil refineries), the prevailing climate can affect the construction programme. Site assembly for large sites using a number of contractors can lead to problems with labour relations.

Unlike packaged firetube boilers, shop-assembled water-tube boilers will in most cases be transported without firing equipment, valves, mountings, instruments and controls, due to the fact that these will occupy transport space that can be more usefully used for boiler surfaces, hence the preference for the term 'shop-assembled' boiler.

Shop-assembled water-tube boilers are available from most water-tube boiler manufacturers as a pre-engineered standardised product range. The output capability depends very much upon the works facilities available, their location and the transport limitations applicable to each manufacturer. Outputs up to 280 MW (101 kg s$^{-1}$) are reported to be available from a single oil- or gas-fired boiler when built at a site with ready access for marine transport. Steam pressures up to 100 bar and temperatures of 550 °C are also possible. A typical two-drum 'D' type shop-assembled boiler is illustrated in Fig. 2.8; other configurations are also used.

These boilers, designed to make maximum use of transport restrictions,

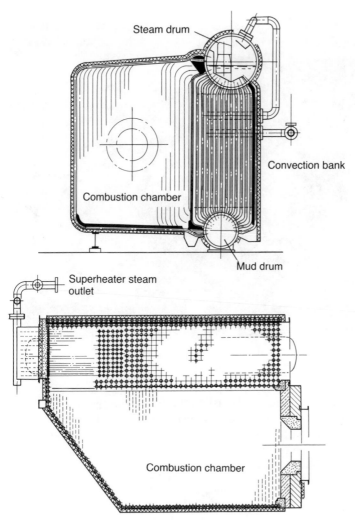

Steam drum

Convection bank

Combustion chamber

Mud drum

Superheater steam outlet

Combustion chamber

**2.8**   'D' type shop-assembled water-tube boiler

are capable of operating at high thermal ratings and also tend to have limited gas-side access and cleaning facilities.

It is, of course, possible to have moderate thermal ratings as required by a particular customer or fuel property simply by selecting a boiler of suitable dimensions from the range available.

They can be site assembled if transport access for a shop-assembled unit is not available but the boiler type is deemed to be suitable for the application in question.

Shop-assembled boilers for firing solid fuels are available, but for much lower maximum outputs than are oil- and gas-fired units. This is due to the limits imposed on thermal ratings by the combustion equipment, the

**2.9** Shop-assembled coal-fired water-tube boiler being lifted at site. (*From Senior Green Limited*)

slagging and fouling properties of the ash in the fuels, and the furnace volume necessary to ensure complete combustion of the fuel. Because of these it is not possible to operate at ratings and temperatures significantly in excess of those used on conventional site-erected boilers. A shop-assembled coal-fired boiler is illustrated by Fig. 2.9.

### 2.4.2 Modular boilers

These are available for a very wide range of applications. The meaning of the term 'modular' is open to broad interpretation. It generally conveys the idea of a unit constructed from very large pre-assembled pieces, the size of which is restricted only by manufacturing facilities and transport limitations to site. Large boilers have for some time been constructed from modules, for example water-wall panels of the combustion chamber, and pre-assembled sections of convection heated surfaces. A typical example of the modern concept of a modular unit is a two-drum natural circulation boiler where the convection bank, consisting of the upper and lower drums and the interconnecting tubes, is despatched to site pre-assembled as one of several modules. Figure 2.10 shows the pressure part modules of a boiler for coal firing on a travelling grate stoker.

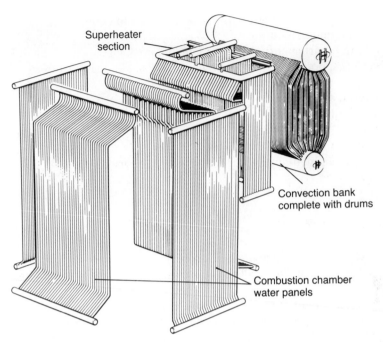

**2.10**  Coal-fired water-tube boiler showing typical modules of construction

The output capability for such units is dictated by the maximum size of convection bank module that can be transported.

### 2.4.3  Site- or field-assembled boilers

These are boilers that are too large for shop assembly or for construction where transport and site restrictions prevent it being carried out. They usually require specialist services on site for functions such as the expanding of tubes or the welding of pressure parts. A typical design of boiler where a significant amount of site assembly is required is illustrated by Fig. 2.5.

## 2.5  Specification of boiler plant

When selecting the unit output and number of boilers to be used for a particular application, the following factors must be considered:

- capital cost;
- limitations of each design available;
- steam pressure;

- steam temperature and the necessity, if any, for this to be maintained constant over a range of output;
- output required, particularly any minimum and peak loads;
- loading patterns, both short- and long-term, and single, two- or three-shift (that is 8-, 16- or 24-hour operation per day);
- standby capacity;
- type of fuel – solid, liquid or gaseous;
- fuel properties;
- properties of ash or non-combustibles;
- type of firing equipment available for the fuels to be used (Ch. 14); and
- site limitations.

### 2.5.1 Operating conditions

For a manufacturer to recommend a unit from his range of products that is most suitable for a specific duty, all the above information should be made available to him.

As a major factor is initial plant cost, there is a requirement to install the minimum spare generating capacity necessary to keep total cost to a minimum and also to select the type of boiler having minimum capital cost while meeting all other requirements. When the steam is used for process requirements the operating conditions of the boiler will be controlled by the highest pressure and temperature required by the consumers. When used for combined process and power or just power generation the conditions will be established by the requirements of the steam turbine.

Where there is freedom to select the pressure and temperature at which the boiler is to operate, discussions should be held with the plant suppliers to determine the pressure limitations of the various designs available. This may prevent steam conditions being selected just beyond the capability of a particular type of boiler, and a more costly boiler than necessary being purchased.

In establishing the boiler operating parameters, care must be taken to allow for the pressure and temperature drop in the pipework between the steam consumer and the boiler. Failure to do this may result in the equipment that is being supplied with steam from the boiler not being capable of achieving its design performance. Even when the consumers require only saturated steam, long runs of pipework can result in wet steam being delivered to them. In such cases modest superheat at the boiler may be desirable.

The steam temperature obtainable from a particular boiler of the type normally used for industrial applications and having a convection superheater will fall with reducing output, see Fig. 2.11 curve (a). The requirement to maintain the temperature constant over a range of boiler outputs therefore necessitates the inclusion of a superheater larger than is

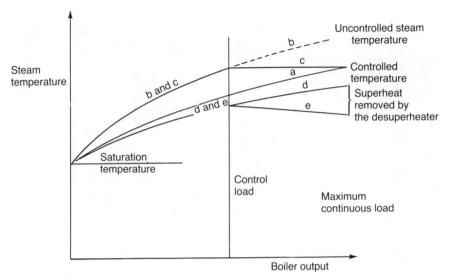

**2.11**  Typical characteristic of a convection superheater:
curve *a* for a superheater that gives the desired steam temperature at full
load only; and curve *b* enables the desired steam temperature to be achieved
from control load upwards, by use of interstage desuperheater giving curve *d*
at primary outlet, and curve *e* at secondary inlet

required to meet the requirement at full load. This will give a steam
temperature higher than necessary at all loads above the minimum or
control load from which it is required to maintain the temperature, see Fig.
2.11 curve (*b*) and, depending upon the temperature required, this may
necessitate the use of some form of steam temperature control equipment
to limit the outlet temperature to that required by the steam consumers
(Fig. 2.11 curve (*c*)).

Steam temperature control, consisting of some form of direct-contact or
non-contact heat exchange, will usually be positioned within the
superheater on the steam side. This necessitates there being at least two
sections called the primary (low-temperature) superheater and the
secondary (high-temperature) superheater (Fig. 2.12) with additional
headers to connect the superheater to the steam temperature controller.
Figure 2.11 curve (*d*) illustrates the temperature of the steam leaving the
primary section and curve (*e*) the temperature of the steam entering the
secondary section after the desuperheater.

It is possible with some designs of boiler to bypass some of the flue gases
around the superheater and hence to reduce the heat transfer and
consequently the steam temperature achieved. This form of control is not
particularly successful due to the need for high-temperature dampers with
which it can be difficult to achieve tight shut-off, resulting in gas leakage at
low loads when bypass is not required.

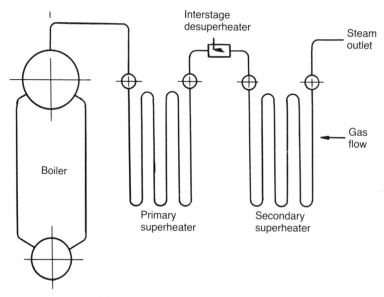

**2.12**   Diagram of the components of a two-stage superheater

Both the larger superheater and steam temperature controls incur increased capital cost, and a constant steam temperature should not therefore be called for unless essential for the steam consumers, for example steam turbines operating at temperatures above about 400 °C.

Steam temperature control devices are often called desuperheaters or attemperators.

### 2.5.2   Number and output of boilers

Determining what number and output of boilers to use is probably the most difficult decision to make when establishing design parameters because it necessitates considering the maximum and minimum outputs required over a complete plant cycle, which may be a day, a week or a year. The periods over which loads occur, plant availability, maintenance periods and the types and costs of boilers available for the steam conditions required must be considered. Loading patterns vary from a continuous load over 365 days per year, as may be required by a continuously operating process plant, down to a heating load which is seasonal and probably non-existent in the summer months. Any future increase in plant capacity should also be considered. It can be beneficial to plot a diagram of the load pattern anticipated.

The first consideration is that of minimum standby requirements. All boilers have to be shut down for inspection at least every 14–26 months for insurance purposes, depending upon the type of boiler and fuel. Shut-down is probably more frequent for cleaning, maintenance and repair. If it is essential for the full steam output to be always available 365 days per year then at least one standby boiler must be included in the system. If it is not possible to tolerate even a short-term reduction in steam flow, it may be necessary to have two standby boilers to ensure that a spare is available to cover unforeseen outages while one is down for maintenance and inspection. This latter period can be extensive if a number of boilers are used and the maintenance periods are all additive. For example, if two weeks per boiler are required for maintenance and a plant consists of six boilers, the minimum period with one boiler out of commission could be twelve weeks per year.

Complete factory shut-down for holidays can be used for maintenance periods, but this necessitates very good planning of the shut-down programme and anticipation of material and labour requirements during the maintenance period.

Where a standby system is required and there is no practical limit to the output available from the type of boiler selected, some of the possible choices for the number of boilers to use are given in Table 2.2. Generally, because small output boilers cost more per unit of output than do large ones, as illustrated by Fig. 2.13, two 100% capacity boilers are likely to give the optimum arrangement, particularly when the complete plant including fuel handling, pipework and civil works is included.

The type of boiler selected can also influence the extent of necessary standby. The boilers can, by choice, be designed with liberal ratings, good on-load cleaning facilities and wide tube spacings to give a high availability, that is the ability to operate for long periods without it being necessary to shut them down for cleaning and maintenance. To incorporate these features does, of course, increase the capital cost.

The conclusion to be drawn, therefore, is that if highly rated shop-assembled boilers are used for essential services where continuity of supply is essential, it is wise to install liberal standby capacity.

Where firetube boilers are capable of meeting the pressure and

**2.2**   Options for number of boilers

| Total number installed | Number on standby | Capacity of each as a percentage of total steam flow | Total installed capacity as a percentage of total steam flow |
|---|---|---|---|
| 2 | 1 | 100 | 200 |
| 3 | 1 | 50 | 150 |
| 4 | 1 | 33 | 133 |
| 5 | 1 | 25 | 125 |

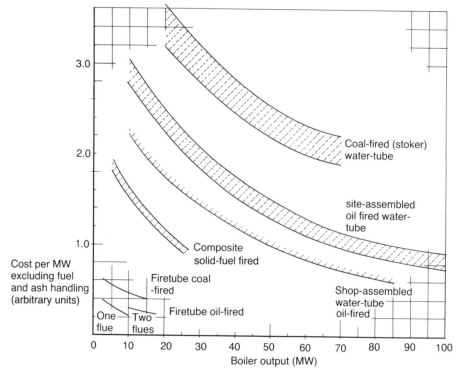

**2.13** Typical ratio of boiler costs

temperature requirements of an installation, even though the output capability per boiler is less than the water-tube type, the price comparison between the two is such that a greater number of firetube boilers could provide an economical solution (Fig. 2.13).

Where the load is seasonal, as with space heating or in a seasonal process plant such as a sugar producing factory, where the plant may be shut down for some weeks (say, 20 plus), it may be possible to run without standby if shut-down for cleaning will not be necessary during the on-load period.

Where an installation has a continuous demand plus a significant seasonal load, by careful boiler size selection the period for which the seasonal load is not required can be used for maintenance and repair, thereby reducing or eliminating the need for standby capacity. Short-term peak loads of a few hours duration and of, say, up to about 10% of the normal maximum load, may be met by running the boilers at overload by utilising the margins in the auxiliaries such as fans and burners. This keeps capital investment at a minimum.

The minimum load required from a boiler plant must be considered because, if this is a very small proportion of the total installed capacity for a significant period and few boilers are used, it may not be practical to operate a single boiler at such a low load either automatically or economically. It is not reasonable to operate a boiler for extended periods at loads below about 20% of the maximum continuous output, that is with a turndown of 5 to 1, because of control problems and reduced efficiency. Indeed, many consider 3 to 1 to be the desirable practical limit. It may therefore be necessary to consider a smaller boiler to meet the low load requirement. If, under such conditions the steam pressure requirements are less stringent, a boiler having a lower operating pressure and hence cheaper design may be applied. An example of such a situation could be a space-heating requirement for which the major steam load is shut down over a holiday period. A combined installation of water-tube and firetube boilers could therefore provide an economical solution. Another example is in oil-fired power stations, where the considerable energy needed at relatively low steam pressure to heat the oil in storage, and additionally to the temperature required at the burners of the power boilers, is often provided by auxiliary firetube boilers, the main boilers of course being water-tube. In this case, there is the added advantage that less stringent feedwater treatment is required for the auxiliary units.

### 2.5.3 Fuels

Assuming that no cheap source of waste fuel or combustible by-product of the plant is available, the choice of fuel is usually one of the fossil fuels, coal, oil, or natural gas, or a combination of these. For a given set of conditions coal-fired boilers are the most expensive and gas-fired the least expensive (Fig. 2.13).

As we have seen over the recent past in the UK, the economics of using a particular fuel can vary for a variety of reasons. In other parts of the world indigenous supplies may be the controlling factor, e.g. oil- or natural-gas firing in oil-producing countries.

The fuel or fuels selected will inevitably dictate the configuration of the boiler heating surfaces from gas-side fouling considerations. Natural gas is a clean fuel and is normally free of compounds such as those of sulphur which could cause corrosion and deposits, see Chapter 10. This enables compact heating surface arrangements, with perhaps no on-load cleaning facilities to be used, giving minimum cost.

Coal is usually the most problematical of the fossil fuels as far as boiler design is concerned. The properties of the fuel, notably the volatile and ash contents together with the fusion characteristics of the ash, influence the size of the furnace, tube spacing and tube arrangement required to avoid slagging and to reduce the effects of fouling of the tubes.

Fuel oils can vary from clean premium fuels which are almost equivalent to natural gas to the very heavy residual fuels with relatively high ash, asphaltenes, vanadium and sulphur contents. The impurities in residual oils can result in heavy deposits on the boiler heated surfaces if the necessary provision is not made in their design, they can also expose the surfaces in contact with the gases to high- and low-temperature corrosion (Ch. 10).

If the boiler is to be provided with the means of firing alternative fuels to cover for a failure in the supply of the main fuel, the design has to make provision for the one with the most difficult combustion characteristics and fouling properties. While it may be an operational advantage to have a boiler plant capable of firing any fuel in an emergency, this will inevitably result in a very high capital cost and more sophisticated control equipment. For example, when a boiler designed for coal firing is fired with oil, the final steam temperature will be lower than that achieved when firing with coal. This is due to the lower gas flow with oil firing compared with that with coal firing (approximately 80%). This carries less heat from the furnace, requiring a greater temperature drop in the superheater for a given heat transfer. The achievement of the desired steam temperature when firing oil will therefore necessitate a larger superheater than is required with coal firing, (see Section 5.6), but tube spacings suitable for coal firing will still need to be incorporated. This will result in the necessity for steam temperature control equipment with a high capacity which, if of the interstage spray type. (Fig. 2.12), will result in a low steam flow through the section preceding the control equipment, thereby affecting the tube metal temperatures and thicknesses, see Chapter 5.

With solid fuels, there are a number of methods of firing, each of which is most suitable for given fuel properties, i.e. stokers, pulverised fuel firing and fluidised bed which are all discussed in Chapter 14. For a boiler to be able to fire a wide range of fuels may mean some compromise may have to be made, and this may result in unsatisfactory combustion in certain cases.

In addition to the effects upon the boiler, the ability to fire multiple fuels necessitates the inclusion of handling and storage equipment for each and will influence the fuel price that can be negotiated with the supplier.

An example of a situation in which dual fuel facilities can assist in reducing fuel costs is the case where natural gas is the main fuel and, if accepted on an interruptible basis, a lower cost can be negotiated (Section 4.1). This does, however, necessitate having a supply of alternative fuel, usually oil, immediately available and hence permanently in store.

It is therefore more advantageous for minimum capital cost and simplicity of design for the range of fuels which the plant is to be capable of firing to be kept to a minimum.

### 2.5.4   Space constraints

Site space and access limitations can impose their own restrictions on the number and type of boilers that can be utilised and to some extent the type of fuel to be used. Modern pneumatic coal handling systems are giving more flexibility in the ability to accommodate coal-fired boilers in restricted areas.[4,5]

Boilers fired with natural gas only require the minimum space to accommodate the complete system, because of the absence of fuel-storage facilities.

Factors influencing the boiler efficiency and the effects of feedwater quality are discussed in Chapters 11 and 6 respectively.

## References

1. BS 2790:1986 *Shell Boilers of Welded Construction*, Clause 3.10. British Standards Institution, London.
2. BS 1113:*Design and Manufacture of Watertube Steam Generating Plant (including superheaters, reheaters and steel tube economisers)*. British Standards Institution, London.
3. Trade literature, ME Boilers Limited.
4. Booklet No. 4 *Storage* and No. 5 *Technical Data on Solid Fuel Plant*. Coal and Ash Handling. National Coal Board, London.
5. Kaye, W. G. *Coal and Ash Handling in Industrial Boiler Houses*. Dansk Ingeniorforening, Copenhagen 1981 (Oct.). (NCB Coal Research Establishment, Stoke Orchard, Cheltenham.)

CHAPTER 3

# Chemical aspects of combustion

## 3.0 Introduction

For the purpose of boiler design it is necessary to obtain considerable information about the fuel and its products of combustion. It is our intention in this chapter to describe the combustion calculations from which the quantity of air required for combustion, and the quantity of waste gas produced, its analysis and density can be derived. The calorific values of fuels are also discussed.

As it will be necessary in some of the combustion calculations to determine the density of a gas, this is dealt with first.

## 3.1 Density of gases

The molar mass of a substance is the sum of the terms – atomic mass × number of atoms – for each element in its chemical formula. Thus for carbon monoxide, CO, there is one atom of carbon and one of oxygen, the atomic masses being 12 and 16 respectively.[1] The molar mass is, therefore, $1 \times 12 + 1 \times 16 = 28$.

For carbon dioxide, $CO_2$, there is one atom of carbon and two of oxygen; the molar mass is, therefore, $1 \times 12 + 2 \times 16 = 12 + 32 = 44$. The process is similar for other substances. Molar mass can be expressed in any mass units.

The molar volume of a gas is the volume occupied by the molar mass at stated conditions of temperature and pressure. In SI units the molar mass of any gas in kilograms occupies 22.4 m³ at 0 °C temperature and 1.013 bar pressure (absolute). Taking $CO_2$ as an example; at 0 °C (273 K) and at 1.013 bar (absolute) the density, i.e. mass per unit volume will be

$$\frac{44}{22.4} = 1.9643 \text{ kg m}^{-3}$$

The temperature and pressure of 0 °C and 1.013 bar have often been referred to as standard temperature and pressure (STP). The term normal temperature and pressure (NTP) has also been used to imply somewhat more normal atmospheric conditions. There are no universally accepted definitions of STP and NTP; the actual pressure and temperature conditions at which the density of a gas is given should, therefore, always be stated to avoid ambiguity.

Where the actual temperature and pressure of the gas vary from 0 °C and 1.013 bar the density at those conditions is calculated by applying the gas laws, i.e. using the equation

$$PV = WRT$$

where:  $P$  is the pressure (absolute);
   $V$  is the volume;
   $W$  is the mass of gas occupying volume $V$;
   $T$  is the temperature in K; and
   $R$  is the universal gas constant.
   (All in self-consistent units.)

$$\text{Density} = \frac{W}{V} = \frac{P}{RT}$$

$$\therefore \text{Density}_1 = \frac{P_1}{RT_1} \tag{3.1}$$

$$\text{Density}_2 = \frac{P_2}{RT_2} \tag{3.2}$$

Therefore, from [3.1] and [3.2],

$$R = \frac{P_1}{T_1 \, \text{Density}_1} = \frac{P_2}{T_2 \, \text{Density}_2}$$

From which

$$\text{Density}_2 = \text{Density}_1 \times \frac{T_1}{T_2} \times \frac{P_2}{P_1} \tag{3.3}$$

Using $CO_2$ as the example:
Density at $t$ °C and $p$ bar will be

$$1.9643 \times \frac{273}{273+t} \times \frac{p}{1.013} \, \text{kg m}^{-3}$$

## 3.2   Density of gaseous mixtures

The density of a mixture of gases can be readily calculated by first determining the molar mass of the mixture and then applying the principle given in section 3.1. As an example, natural gas of the analysis given in Table 3.1 is used.

**3.1** Determination of analysis by mass from an analysis by volume for natural gas

| Component | Formula | Percentage by volume $A$ | Proportion by volume $B = \dfrac{A}{100}$ | Molecular mass $C$ | Proportion by mass $D = B \times C$ | Percentage by mass $E = \dfrac{D \times 100}{\text{Total of } D}$ | Proportion by mass $F = \dfrac{D}{\text{Total of } D}$ |
|---|---|---|---|---|---|---|---|
| Methane | $CH_4$ | 94.5 | 0.945 | 16 | 15.120 | 89.224 | 0.8922 |
| Ethane | $C_2H_6$ | 2.9 | 0.029 | 30 | 0.870 | 5.134 | 0.0513 |
| Propane | $C_3H_8$ | 0.4 | 0.004 | 44 | 0.176 | 1.039 | 0.0104 |
| Butane | $C_4H_{10}$ | 0.2 | 0.002 | 58 | 0.116 | 0.685 | 0.007 |
| Pentane | $C_5H_{12}$ | 0.1 | 0.001 | 72 | 0.072 | 0.425 | 0.0042 |
| Hydrogen Sulphide | $H_2S$ | 0.2 | 0.002 | 34 | 0.068 | 0.401 | 0.0040 |
| Carbon Dioxide | $CO_2$ | 0.3 | 0.003 | 44 | 0.132 | 0.779 | 0.0078 |
| Nitrogen | $N_2$ | 1.4 | 0.014 | 28 | 0.392 | 2.313 | 0.0231 |
| Totals | | 100.0 | 1.000 | | 16.946 | 100.0 | 1.000 |

Molar mass of mixture = 16.946

The molar mass of the mixture is obtained by multiplying the proportion by volume of each constituent by its molar mass and summing the results. A summary is given in Table 3.1, in which column $B$ × column $C$ gives column $D$, and the total for column $D$ gives the molar mass of the mixture of gases, that is 16.946. The density of the mixture (natural gas) at 0 °C and 1.013 bar, therefore, becomes $\dfrac{16.946}{22.4} = 0.7565$ kg m$^{-3}$.

As an alternative, if values for the density of each constituent gas are readily available, the proportion by volume (column $B$) of each constituent can be multiplied by its density and the results summed to give the density of the mixture. Care should be taken to ensure that all the density values used are at the same conditions of temperature and pressure when determining the density of the mixture by this method. Conversion to the desired conditions can be carried out as indicated by equation [3.3].

## 3.3 The chemical composition of fuels

The combustible elements present in all fossil fuels, coal, oil and natural gas, are carbon, hydrogen and, in some cases, sulphur. They are not present in elementary form but are combined between themselves and sometimes with other substances. Inert materials such as silica and metallic oxides may also be present.

The analyses of solid and liquid fuels are presented on a mass basis, those of gaseous fuels generally by volume. Analysis for solid fuels can be either 'proximate' or 'ultimate'. The 'proximate analysis' determines moisture, ash, volatiles and fixed carbon contents as a percentage by weight (Table 3.2) and also the calorific value (Section 3.5). The proximate

**3.2**  Proximate analysis of a typical coal (percentage by mass)

| Constituent | As fired | Dry | Dry, ash free |
|---|---|---|---|
| Fixed Carbon | 50.19 | 57.03 | 62.5 |
| Volatiles | 30.11 | 34.22 | 37.5 |
| Ash | 7.7 | 8.75 | — |
| Moisture | 12.0 | — | — |
| TOTAL | 100.0 | 100.0 | 100.0 |
| Gross calorific value kJ kg$^{-1}$ | 27 000 | 30 682 | 33 624 |

analysis gives limited indication of the combustible properties of a fuel.
The ultimate analysis of solid and liquid fuels gives the elements of which
the combustible content of the fuel is composed, along with the ash and
moisture contents (see Table 3.3). The ash can be broken down into its
individual constituents by a separate analysis if required, (Table 10.1). Gas
analyses usually give, on a volume basis, only the constituent gases that
make up the whole.

Ultimate analyses of fuels are required to enable the combustion air
requirements and products of combustion to be determined. Solid fuel
analyses are carried out to British Standard 1016. Liquid fuel analyses are
carried out to the Institute of Petroleum (IP) Standards for petroleum and
its products. Gaseous fuel analyses are usually determined at source and
provided by the fuel supplier.

Solid fuel analyses, particularly those of coal, are specified on a number
of bases; 'as received', 'as fired', 'dry' and 'dry ash-free' or 'dry mineral-
matter free', being typical. The as received and as fired analyses, both
ultimate and proximate, usually only differ in the moisture content, which
may vary during storage and handling, depending upon the climatic
conditions. The dry analysis gives the percentage make-up of the
constituents with no moisture, the dry ash-free gives it with no ash or

**3.3**  Ultimate analyses of typical washed smalls coal and heavy fuel oil
(percentage by mass)

| | Coal | Heavy fuel oil |
|---|---|---|
| Carbon | 66.0% | 85.4% |
| Hydrogen | 4.1% | 11.4% |
| Sulphur | 1.7% | 2.8% |
| Oxygen | 7.2% | 0.1% |
| Nitrogen | 1.3% | 0.1% |
| Moisture | 12.0% | 0.1% |
| Ash | 7.7% | 0.1% |
| Gross calorific value kJ kg$^{-1}$ | 27 000 | 42 900 |

moisture included (see Table 3.2). The dry constituents are obtained by multiplying the as received or as fired constituents by

$$\frac{100}{100 - \text{percentage moisture}}.$$

The dry ash-free constituents are obtained by multiplying the as received or as fired constituents by

$$\frac{100}{100 - (\text{percentage ash} + \text{moisture})}.$$

The ash and mineral matter are not quite the same since some mineral matter will decompose during the heating process used to determine the ash; the ash quantity is, therefore, somewhat less than that of the mineral matter. The dry mineral-matter free analysis is sometimes referred to as 'Parr's basis'.

## 3.4  Combustion calculations

These are based on elementary chemical equations for the reactions with oxygen of each combustible constituent of the fuel.

### 3.4.1  Gaseous fuels

Gaseous fuels may vary between rich liquified petroleum gases comprising a mixture of propane and butane, to lean gases such as blast furnace gas, the combustible content of which is only some 22.0% carbon monoxide with 2.4% hydrogen diluted by a large amount of inert material, carbon dioxide and nitrogen (see Table 15.4 for typical examples). By far the most common gas used in boilers is natural gas. This consists mainly of methane, a typical analysis being shown in Table 3.1. North Sea gas, the basis for Table 3.1 is virtually sulphur free, but a small amount of hydrogen sulphide has been included to illustrate the combustion calculations since sulphur could be a significant constituent in other fuel gases. The chemical reactions of the combustible gases with oxygen are as follows on a volume basis.

- Methane 94.5% by volume

  basic reaction $CH_4 + 2O_2 \rightarrow CO_2 + 2H_2O$

  multiplying through by 0.945, the proportion by volume

  0.945 volumes of $CH_4$ + 1.89 volumes of $O_2$ produce
  0.945 volumes of $CO_2$ + 1.89 volumes of $H_2O$                    [3.4]

- Ethane 2.9%

$C_2H_6 + 3.5O_2 \rightarrow 2CO_2 + 3 H_2O$

multiply through by 0.029

$0.029\ C_2H_6 + 0.1015\ O_2 \rightarrow 0.058\ CO_2 + 0.087\ H_2O$ [3.5]

- Propane 0.4%

  $C_3H_8 + 5\ O_2 \rightarrow 3\ CO_2 + 4\ H_2O$

  multiply through by 0.004

  $0.004\ C_3H_8 + 0.02\ O_2 \rightarrow 0.012\ CO_2 + 0.016\ H_2O$ [3.6]

- Butane 0.2%

  $C_4H_{10} + 6.5O_2 \rightarrow 4\ CO_2 + 5\ H_2O$

  multiply through by 0.002

  $0.002\ C_4H_{10} + 0.013\ O_2 \rightarrow 0.008\ CO_2 + 0.01\ H_2O$ [3.7]

- Pentane 0.1%

  $C_5H_{12} + 8\ O_2 \rightarrow 5\ CO_2 + 6\ H_2O$

  multiply through by 0.001

  $0.001\ C_5H_{12} + 0.008\ O_2 \rightarrow 0.005\ CO_2 + 0.006\ H_2O$ [3.8]

- Hydrogen sulphide 0.2%

  $H_2S + 1.5\ O_2 \rightarrow SO_2 + H_2O$

  multiply through by 0.002

  $0.002\ H_2S + 0.003\ O_2 \rightarrow 0.002\ SO_2 + 0.002\ H_2O$ [3.9]

It should be appreciated that volumes cannot be summed across an equation; this is illustrated in equations [3.5] to [3.9] where the volumes on opposite sides are unequal.

The next step is to summarise equations [3.4] to [3.9] in tabular form to obtain the total stoichiometric quantity of oxygen required (Table 3.4). Allowance is made for the nitrogen in the air supplied and for the necessary excess air added for satisfactory combustion (10% in this case). The products of combustion are then calculated on the wet and dry basis.

In practice the combustion air will contain a small amount of moisture (humidity) which, if it is known, can be calculated and added to the products of combustion along with the excess air in Table 3.4.

### 3.4.2 Products of combustion

Many, but not all, boiler-house instruments for analysing flue gases remove the water vapour resulting from the combustion of hydrogen and from the

## 3.4  Summary of equations 3.4 to 3.9 and analysis of products of combustion

| Equation number | Component gas | Percentage by volume of component | Proportion by volume of component | Oxygen required for stoichiometric combustion, volumes from 3.4–3.9 | Products of combustion with oxygen | | | |
|---|---|---|---|---|---|---|---|---|
| | | | | | Volumes | | | |
| | | | | | $CO_2$ | $H_2O$ | $SO_2$ | $N_2$ |
| 3.4 | $CH_4$ | 94.5 | 0.945 | 1.890 | 0.945 | 1.890 | — | — |
| 3.5 | $C_2H_6$ | 2.9 | 0.029 | 0.102 | 0.058 | 0.087 | — | — |
| 3.6 | $C_3H_8$ | 0.4 | 0.004 | 0.020 | 0.012 | 0.016 | | |
| 3.7 | $C_4H_{10}$ | 0.2 | 0.002 | 0.013 | 0.008 | 0.010 | — | — |
| 3.8 | $C_5H_{12}$ | 0.1 | 0.001 | 0.008 | 0.005 | 0.006 | — | — |
| 3.9 | $H_2S$ | 0.2 | 0.002 | 0.003 | — | 0.002 | 0.002 | — |
| In fuel | $CO_2$ | 0.3 | 0.003 | — | 0.003 | — | — | — |
| In fuel | $N_2$ | 1.4 | 0.014 | — | — | — | — | 0.014 |
| Totals | | 100.0 | 1.000 | 2.036 | 1.031 | 2.011 | 0.002 | 0.014 |

Air consists of 21% by volume of oxygen and 79% by volume of nitrogen.
Stoichiometric air requirements to supply 2.0355 volumes of oxygen are given by:

$2.0355\ O_2 + 7.657\ N_2 = 9.693$ volumes of air

Air requirements including 10% excess air are:

$2.239\ O_2 + 8.423\ N_2 = 10.662$ volumes of air

Total products of combustion (wet) with 10% excess air including nitrogen from the fuel are as follows:

$(2.239–2.0355)\ O_2 + (8.423 + 0.014)\ N_2 + 1.031\ CO_2 + 2.011\ H_2O + 0.002\ SO_2 = 11.685$ volumes/volume of fuel

Total products of combustion (dry) with 10% excess air including nitrogen in the fuel

$(2.239–2.0355)\ O_2 + (8.423 + 0.014)\ N_2 + 1.031\ CO_2 + 0.002\ SO_2 = 9.674$ volumes/volume of fuel

These are summarized in Table 3.5.

moisture, if any, in the fuel. This is achieved by condensation and sometimes by further drying, so that the analysis on the dry basis, i.e. with the water vapour removed, is appropriate. The total volume of dry gases is less than the total volume wet, and hence the percentages of the constituents are different (Table 3.5). It is essential, therefore, that when quoting a flue gas analysis the basis, dry or wet, is stated.

Some oxygen meters do not withdraw a sample of flue gases. They use a zirconia probe inserted into the flue gas duct where the water vapour is still present. The measurement is thus on the wet basis. The mass flow of flue

## 3.5  Products of combustion on wet and dry bases for natural gas

| Product of combustion | Volumes wet | Percentage volume wet | Volumes dry | Percentage volume dry |
|---|---|---|---|---|
| $CO_2$ | 1.031 | 8.823 | 1.031 | 10.657 |
| $H_2O$ | 2.011 | 17.21 | — | — |
| $SO_2$ | 0.002 | 0.017 | 0.002 | 0.021 |
| $N_2$ | 8.437 | 72.20 | 8.437 | 87.213 |
| $O_2$ | 0.204 | 1.75 | 0.204 | 2.109 |
| Total | 11.685 | 100.00 | 9.674 | 100.00 |

## 3.6  Calculation of density of flue gases

| Gas | Molar mass $A$ | % volume wet $B$ (From Table 3.5) | $\dfrac{A \times B}{100}$ | % volume dry $C$ (From Table 3.5) | $\dfrac{A \times C}{100}$ |
|---|---|---|---|---|---|
| $CO_2$ | 44 | 8.823 | 3.88 | 10.657 | 4.689 |
| $H_2O$ | 18 | 17.210 | 3.10 | — | — |
| $SO_2$ | 64 | 0.017 | 0.01 | 0.021 | 0.013 |
| $N_2$ | 28 | 72.200 | 20.22 | 87.213 | 24.420 |
| $O_2$ | 32 | 1.750 | 0.56 | 2.109 | 0.675 |
| Totals | | 100.00 | 27.77 | 100.00 | 29.797 |

Density of wet flue gas $= \dfrac{27.77}{22.4} = 1.24 \text{ kg m}^{-3}$

Density of dry flue gas $= \dfrac{29.797}{22.4} = 1.33 \text{ kg m}^{-3}$

gases for boiler design can be determined by calculating the density as explained in section 3.2 and illustrated in Table 3.6.

### 3.4.3.  Solid and liquid fuels

The analyses of solid and liquid fuels are based on the masses of the combustible substances present as chemical elements, unlike those of gases which are based on volumetric proportions of constituent gases such as methane $CH_4$. Solid fuel will also contain mineral matter and moisture. The former is converted to ash during the combustion process; the latter will evaporate and dilute the products of combustion. Residual liquid fuels, such as heavy fuel oil, will also contain mineral matter and moisture, but to a much lesser degree than most solid fuels, totalling perhaps 0.2% compared with about 20% for a coal of washed smalls grade. Table 3.3 gives typical analyses for such fuels.

To calculate the air required for combustion and the analysis of the products of combustion, the procedure is rather similar to that for gaseous fuels, both for coal and for oil. Coal will be taken for the example, the combustible substances present being carbon, hydrogen and sulphur. It will be seen that oxygen, to a significant extent, appears in the analyses. This is already combined with some of the fuel components and is therefore deducted from the total oxygen required for stoichiometric combustion.

The equations for combustion will now be considered, 40% excess air being appropriate for coal firing in medium-sized boilers (less in large water-tube boilers). For oil firing 17% excess air is normal in medium-sized boilers, less on large water-tube boilers.

Because the fuel analysis is given on a mass basis the combustion calculations will be carried out on a mass basis.

The basic equation for combustion of carbon is

$$C + O_2 = CO_2$$

The molar mass of carbon is 12 and that of oxygen is 32, the mass balance therefore becomes

12 mass units of carbon + 32 mass units of oxygen give 44 mass units of carbon dioxide.

There is 66% carbon in the fuel analysis in Table 3.3, i.e. 0.66 kg of carbon per kg of fuel. Therefore for the fuel in question, the molar masses are multiplied through by 0.66/12, giving that 0.66 kg of carbon requires 1.76 kg oxygen to give 2.42 kg carbon dioxide. The equation becomes

$$0.66 \, C + 1.76 \, O_2 = 2.42 \, CO_2 \tag{3.10}$$

The other combustible constituents are treated in the same manner.

- Hydrogen 4.1%, that is 0.041 kg/kg of fuel

$$2 \, H_2 + O_2 = 2 \, H_2O$$
$$4 \quad + 32 = 36$$

multiply through by $\dfrac{0.041}{4}$

$$0.041 \, H_2 + 0.328 \, O_2 = 0.369 \, H_2O \tag{3.11}$$

- Sulphur 1.7%, that is 0.017 kg/kg of fuel

$$S \quad + O_2 = SO_2$$
$$32 \quad + 32 = 64$$

multiply through by $\dfrac{0.017}{32}$

$$0.017 \, S + 0.017 \, O_2 = 0.034 \, SO_2 \tag{3.12}$$

The results are now tabulated in Table 3.7 making due allowance when calculating the oxygen for stoichiometric combustion for the oxygen already in the fuel. Forty per cent excess air is added and the flue gas analysis produced on a mass balance.

In this instance, with the calculations being carried out on a mass basis, it is possible to check the results, for example:

mass of total air supplied + mass of ash free fuel = mass of products of combustion

12.374 kg air/kg fuel + 0.923 kg ash-free fuel = 13.297 kg flue gases as fired

This enables the gas mass flow through the boiler to be readily established for heat transfer calculations.

**3.7**  Analysis of products of combustion from coal by mass

| Fuel constituent | % present in fuel by mass | Oxygen required for stoichiometric combustion kg/kg fuel | Products of combustion with oxygen kg gas/kg fuel | | | |
|---|---|---|---|---|---|---|
| | | | $CO_2$ | $H_2O$ | $SO_2$ | $N_2$ |
| C | 66.0 | 1.760 | 2.42 | — | — | — |
| H | 4.1 | 0.328 | — | 0.369 | — | — |
| S | 1.7 | 0.017 | — | — | 0.034 | — |
| $O_2$ | 7.2 | −0.072 | — | — | — | — |
| $N_2$ | 1.3 | | — | — | — | 0.013 |
| $H_2O$ | 12.0 | | — | 0.120 | — | — |
| Ash | 7.7 | | — | — | — | — |
| Totals | | 2.033 | 2.42 | 0.489 | 0.034 | 0.013 |

Air consists of 23% by weight of oxygen and 77% by weight of nitrogen. Stoichiometric air requirements to supply 2.033 kg of oxygen are given by

$2.033 \, O_2 + 6.806 \, N_2 = 8.839$ kg air

Air requirements including 40% excess air are

$2.846 \, O_2 + 9.528 \, N_2 = 12.374$

Total products of combustion (wet) with 40% excess air and including nitrogen in the fuel are as follows

$(2.846 - 2.033) \, O_2 + (9.528 + 0.013) \, N_2 + 2.42 \, CO_2 + 0.489 \, H_2O + 0.034 \, SO_2 = 13.297$ kg gas/kg fuel

Total products of combustion (dry) with 40% excess air and including nitrogen in the fuel are as follows

$(2.846 - 2.033) \, O_2 + (9.528 + 0.013) \, N_2 + 2.42 \, CO_2 + 0.034 \, SO_2 = 12.808$ kg gas/kg fuel

These are summarised on Table 3.8

Calculations for fuel gases can be carried out on a mass basis by first converting the analysis to a mass basis as column $F$ of Table 3.1.

### 3.4.4  Solid and liquid fuels – products of combustion

To convert these from a mass basis to a volume basis, the reverse of the process given in section 3.2 is carried out where the proportion by mass of each constituent is divided by its molecular mass, giving a volumetric ratio from which the percentage by volume can be determined, both wet and dry. The results are summarised in Table 3.8.

When operating a boiler it is useful to know the relationship between the oxygen and carbon dioxide contents in the products of combustion and the

## 3.8   Analysis of products of combustion of coal

| Product of Combustion | Molecular Mass $A$ | Mass dry kg/kg fuel $B$ | Proportion by volume $\frac{B}{A} = C$ | Percentage by volume dry | Mass wet kg/kg fuel $B$ | Proportion by volume $\frac{B}{A} = D$ | Percentage by volume wet |
|---|---|---|---|---|---|---|---|
| CO$_2$ | 44 | 2.420 | 0.0550 | 13.046 | 2.420 | 0.0550 | 12.26 |
| H$_2$O | 18 | — | — | — | 0.489 | 0.0270 | 6.02 |
| SO$_2$ | 64 | 0.034 | 0.0005 | 0.119 | 0.034 | 0.0005 | 0.11 |
| N$_2$ | 28 | 9.541 | 0.3407 | 80.81 | 9.541 | 0.3407 | 75.95 |
| O$_2$ | 32 | 0.813 | 0.0254 | 6.025 | 0.813 | 0.0254 | 5.66 |
| Totals | | 12.808 | 0.4216 | 100 | 13.297 | 0.4486 | 100.0 |

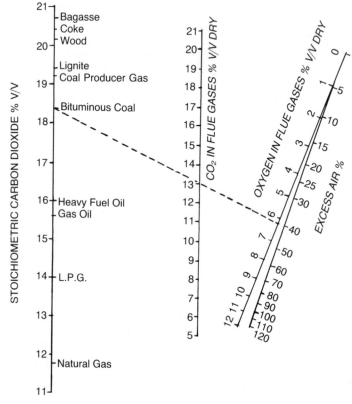

**3.1**   Relationship between excess air, carbon dioxide and oxygen content of flue gases for various fuels

excess air for the fuel being fired. For this purpose, Fig. 3.1, or graphs developed from a series of calculations with varying excess air quantities for each fuel, can be used.

Alternatively, it may be convenient to use a formula for this purpose; this is

$$VCO_2 = \left(1 - \frac{VO_2}{21}\right) \bar{V}CO_2 \qquad\qquad [3.13]$$

where:   $VCO_2$  is the percentage by volume of carbon dioxide in the flue
gas;
$VO_2$   is the percentage by volume of oxygen in the flue gas; and
$\bar{V}CO_2$  is the stoichiometric percentage by volume of carbon
dioxide for the fuel.

It is worth noting that the density of the products of combustion of
natural gas is 1.24 kg m$^{-3}$ compared with that for the products of
combustion of coal which is 1.323 kg m$^{-3}$. For heavy fuel oil a figure
midway between these will be appropriate. For air, the density is
1.29 kg m$^{-3}$; for approximate assessment, therefore, the density of the
products of combustion from commercial fuels can be taken as that for air
without serious error.

## 3.5   Calorific value (specific energy)

The calorific value of a fuel is the quantity of heat produced by the
combustion of unit mass of fuel under specified conditions of temperature
and pressure. It can be determined by burning a measured quantity of fuel
in a calorimeter in which all the heat released can be measured and
accounted for. A more complete description of the process can be found
elsewhere.[2] Realistic figures for calorific value can be obtained from the
fuel supplier as 'declared' values, or by submitting for analysis a sample
taken at the point of usage, which, particularly with coal because of
moisture variation, is probably the best approach. For distributed natural
gas, and bulk liquified petroleum gas, the declared value is adequate. With
oil there is a strong correlation[3] between calorific value, density and
sulphur content which is now often used in preference to a calorimetric
determination.

It is important to appreciate the difference between gross and net
calorific values. The gross calorific value is the total energy released,
whereas the net value discounts the latent heat of the water vapour in the
products of combustion as being not available for useful heat transfer.
With some modern boilers this is not true, as some condensation is
intended to take place. When quoting a calorific value of a fuel or the
efficiency of a boiler, the basis, gross or net, should always be stated as
otherwise confusion can arise. Since the net calorific value is the smaller
figure (in Europe it is called the 'lower calorific value') the efficiency of a
boiler, i.e. the percentage of the heat supplied which is converted to useful
energy at a given output is a higher figure if based on the net or lower
rather than the gross or higher calorific value. The basis 'gross' or 'net' is
sometimes omitted deliberately by the unscrupulous to deceive the unwary!

Sometimes a fuel will be used in a boiler and the calorific value is unknown. If its analysis, volumetric for a gas, by mass for a solid or liquid, is known, then the calorific value can be estimated. For a gaseous fuel this can be done in principle, by multiplying the percentage of each combustible component present by its calorific value which can be obtained from an appropriate reference book.[4] The products are summed, and the total divided by 100. In practice, it is not quite as simple as that, as a compressibility factor needs to be introduced, the reader is advised to study the procedure described elsewhere.[5]

For fossil fuels, coal and oil the formula due to Dulong may be used when the oxygen content is less than 10%. This gives

$$Q_g = 337\ C + 1442\left(H - \frac{O}{8}\right) + 93\ S \qquad [3.14]$$

where: $Q_g$ is the gross calorific value (kJ kg$^{-1}$);
     $C$ is the percentage carbon, as fired (%);
     $H$ is the percentage hydrogen, as fired (%);
     $O$ is the percentage oxygen, as fired (%); and
     $S$ is the percentage sulphur, as fired (%)
     (the percentages being on a weight basis).

For biomass fuels and refuse, where the oxygen content exceeds 10%, the formula due to Vondracek should be used. This gives

$$Q_g = C\left(373 - 26.1\frac{C_1}{100}\right) + 1130\left(H - \frac{O}{10}\right) + 105\ S \qquad [3.15]$$

Where: $C_1$ is the percentage carbon on the dry ash-free basis, the other symbols, including $C$, being as for equation [3.14].

Both of these formulae appear to give high results, by about 5%. In order to provide some idea of the range of gross calorific values that may be encountered, the following are some typical figures on the as fired basis.

Coal, washed smalls  27 000 kJ kg$^{-1}$
Wood  16 000 kJ kg$^{-1}$ (20% moisture)
Heavy fuel oil  42 900 kJ kg$^{-1}$
Gas oil  45 000 kJ kg$^{-1}$
Natural gas  38 MJ m$^{-3}$

These are indicative only, more precise figures can be found elsewhere for solid,[6] liquid[7] and gaseous[8] fuels.

# References

1. Rose, J. W. and Cooper J. R.
   *Technical Data on Fuel* (7th edn),
   1977. British National Committee,
   World Energy Conference London,
   p. 153.
2. Ibid., p. 127.
3. BS 2869:1983 *Fuel Oils for Oil
   Engines and Burners for non-Marine
   use*. British Standards Institution,
   London, Fig. 4, p. 9.
4. Rose, J. W. and Cooper J. R.,
   *Technical Data on Fuel* (7th edn),
   1977. British National Committee,
   World Energy Conference, London,
   p. 267.
5. Ibid. p. 132.
6. Ibid. pp. 302–4.
7. Ibid. p. 283.
8. Ibid. pp. 268, 269.

# CHAPTER 4

# Physical aspects of combustion

## 4.0 Introduction

It was explained in the previous chapter that combustion is a chemical process involving the combination of carbon and hydrogen compounds with oxygen and diluents, the latter being mostly nitrogen.

Thorough mixing of the reacting substances is needed to ensure complete combustion and some excess oxidant, usually air, is needed due to the difficulty of achieving perfect mixing with practical combustion appliances. There are other factors: time, temperature and turbulence, the 'three T's'; which also need to be considered, these having a dominant influence on the success or otherwise of a combustion system. These are physical factors and it will be appreciated that the methods used to satisfy them may differ between gaseous, liquid and solid fuels.

The simplest case is that of the combustion of a gaseous fuel; mixing a gas with a gas is easier than mixing a gas with a liquid or a solid. In the case of liquid fuels these can be brought into a near-gaseous state by finely atomising them, the droplets being suspended in air. The same can be done with some solid fuels, e.g. coal, which can be finely ground or pulverised. In many cases, however, coal is burned as larger particles on grates where the conditions are completely different.

It is the purpose of this chapter to explain these physical aspects and to show how they are incorporated into practical combustion appliances for industrial boilers.

## 4.1 Gaseous and liquid fuels

These will be discussed together, since the combustion appliances involved have many features in common. Indeed, many burners are designed for both types of fuel in order to take advantage of the interruptible tariffs offered by the gas industry, whereby very favourable terms are available in return for the ability to cease demand for gas at short notice. This means that a switch to an alternative fuel, usually oil, must be made. Most industrial gas-fired boilers are equipped with these dual fuel burners.

The requirements for the satisfactory combustion of gaseous, liquid and pulverised solid fuel are as follows.

1.  The fuel/air mixture should be readily ignitable.
2.  The resultant flame shall be perfectly stable under all operating conditions of the boiler.
3.  The flame shall be wholly contained within the furnace volume.
4.  Specified limitations to the emission of unburned gases and particulates shall be met.
5.  Combustion shall be completed to the requirements of (4) with a minimum of excess air.

### 4.1.1   Ignition

In order to ignite a fuel/air mixture energy is needed. This is minimal in the case of gas, but with oil the droplets need to be partially vapourised, which will require more energy. With pulverised fuel (PF) still more energy will be needed. All industrial gas-fired and oil-fired boilers are now fully automatic in operation, energy being provided in the first place by an electric spark at about 11 kV. This ignites a gas pilot flame which in turn ignites the main flame, providing amplified energy for this purpose. The amount of energy needed varies between different fuels, with local fuel/air ratio, and with local mixture velocity.[1] The fuel/air ratio needs to be between the limits of flammability of the mixture, least energy being needed when it is near stoichiometric (i.e. the minimum amount of air needed theoretically to burn the fuel; see equations [3.4] to [3.9] Ch. 3). The energy needed increases with mixture velocity; it is for this reason that pilot flames are designed for low velocity and that the main flame is set to ignite at minimum firing rates. It is also important that the mixing is sufficiently rapid to ensure that, at least local to the ignition source, near stoichiometric conditions are established.

### 4.1.2   Flame stability

Once the flame has been established, and proved to have been established, the ignition source is extinguished. The main flame must then be entirely stable and reliable. In some earlier types of burner this was not always the case, flames sometimes oscillated quite violently ('buffeting'), a condition which is not acceptable for the operation of any boiler, particularly large ones.

The flame is not attached to the burner throat, but is stabilised some small distance away, the distance being determined by the burning velocity of the fuel/air mixture relative to the rate of flow of mixture at the burner mouth, and by the space and time needed for the mixture to assume near-stoichiometric proportions. It follows that mixing of fuel and air must be rapid, and that an environmental temperature exceeding the ignition temperature of the fuel must be maintained in the region of the flame front.

With gas, the air and fuel can be mixed, at least partially, before arrival at the burner. This is known as pre-mixing, but is never used in boiler practice owing to the liability of the mixture to ignite explosively back into the burner throat under low fire conditions. The alternative is to mix the fuel and air very near to the burner throat, thus avoiding the problem. Atomised oil, from its nature, must be mixed with air downstream from the burner throat, otherwise oil will be precipitated on the burner parts, becoming, in effect, de-atomised. In order to accomplish mixing, the air is given turbulence by causing it to swirl. Axial swirl may be induced by vanes upstream of the burner throat, as illustrated in Fig. 4.1(a). Where gas is used this may also be swirled sometimes with gas and air, in the opposite direction. Axially swirled burners are generally used in the larger water-tube boilers but not often in firetube boilers. The alternative method of inducing swirl is to pass the air over a bluff body, usually in the form of a perforated disc fixed around the oil atomiser, as illustrated in Fig. 4.1(b). This causes the air to eddy strongly after it has passed through the annulus between the bluff body and the burner throat, in a toroidal manner, so mixing with the fuel, be it gas or oil. The more intense the swirl, the greater is the mixing rate, but the pressure needed to cause air flow through the burner increases and, generally, the more compact is the flame. The pressure loss through the burner is, particularly in the larger boilers, of economic importance; it should be minimised to an extent commensurate with obtaining the required flame characteristics. The mass of air is about sixteen times the mass of fuel, so that it is the air pattern rather than that of the fuel which dominates the mixing process.

At low turndown, i.e. when burners are firing below their maximum output, the amount of air, and hence the turbulence, decreases and the effectiveness of mixing likewise decreases. To compensate for this, more excess air is used at low turndown, partly to make good the deficiency in flow rate and hence to increase the turbulence, and partly to provide a greater proportion of oxidant to consume the products of partial combustion which can arise through less effective mixing; a method for overcoming this problem is described later in this chapter.

Induced swirl has another important effect: it causes hot gases from the

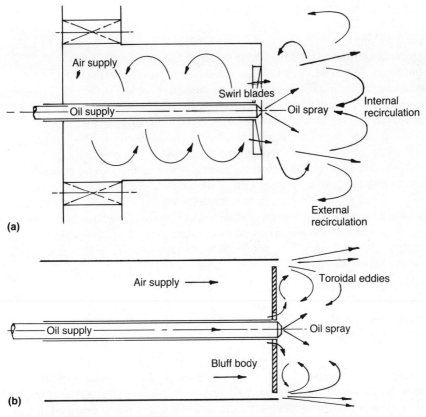

**4.1**   Principles of flame stabilisation
(*a*)  using axial swirl, and
(*b*)  using a bluff body

flame to be recirculated back to the ignition zone, thus raising the
temperature in this region, particularly that of the refractory material
surrounding the burner throat, the quarl. This increases the burning rate.
Excessive recirculation can, however, have an adverse effect on flame
stability. First, the recirculated gases need to be driven forward again, their
added volume increasing the forward velocity of the flame, which tends to
act against the increase in burning rate. Second, much of the recirculated
gases consists of inert products of combustion as well as substances in the
process of combustion. The former tends to decrease the burning rate, the
latter to enhance it. Recirculation therefore is a matter for optimisation,
which needs skill and experience to arrive at the right value. Typical
recirculation patterns are illustrated by Fig. 4.1(*a*) and 4.1(*b*). For a more
detailed study of the aerodynamics affecting flames the reader is referred
elsewhere.[2]

### 4.1.3 Containment of the flame within the furnace

This is an important consideration in that it is the furnace, and the furnace only, that is intended for the space in which combustion is to be completed. Extension of the flame beyond the furnace will lead to problems, including damage to the pressure parts of the boiler, deposits on heated surfaces and chilling of reacting gases which in turn causes incomplete combustion. It follows therefore that there must be mutual optimisation between furnace geometry and flame pattern. This is another area where experience is needed. The larger the furnace volume the greater is the time available for the combustion reactions to be completed, and the less likely it is for problems of incomplete combustion to occur.

To provide some guidance in this matter, current practice suggests that, both for gas and heavy fuel oil, the following values for volumetric heat release rate, based on the gross calorific value of the fuel, are typical maxima, although they have on occasions been much exceeded, albeit with associated problems.

| | |
|---|---|
| Firetube boilers | $1.9$ MW m$^{-3}$ |
| Water-tube boilers, site erected | $0.3$–$0.4$ MW m$^{-3}$ |
| Water-tube boilers, factory assembled | $0.6$–$0.8$ MW m$^{-3}$ |
| (See also Table 13.3) | |

In firetube boilers, the standards of construction[3] impose limitations on heat input to furnaces related to their diameters. The maximum input, based on the net calorific value is 12 MW for a furnace of internal diameter 1.4 m up to a maximum permitted diameter of 1.8 m when firing oil or gas. Furnaces of smaller diameter are permitted less heat input (see Fig. 12.1). It is also mandatory in the standards that 'combustion shall be completed within the furnace'.

### 4.1.4 Unburned particles and gases

Figure 4.2 shows the relationship between particulate emission (stack solids) and volumetric heat release in a firetube boiler burning medium fuel oil (Class F, BS 2869). This was experimentally derived and indicates that the emission started to escalate beyond a thermal input of $2.0$ MW m$^{-3}$ (modified burner). The picture would be worse with heavy fuel oil (Class G, BS 2869), especially with oils of high asphaltene content. For this reason the figures quoted above should be taken as maxima rather than as recommended figures. There should be good consultation between the boiler designer, the burner designer and the fuel supplier, to arrive at a satisfactory figure, which could be nearer $1.5$ MW m$^{-3}$ for heavy fuel oil in firetube boilers.

It will be seen that for the original burner, fitted to the same boiler, the

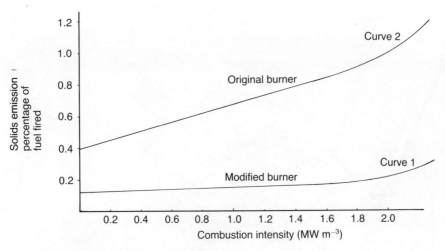

**4.2**   Solids emission with medium fuel oil

emission was much greater and always exceeded the present legal limit in the UK (0.4% of the fuel fired). This was an older type of burner using the same atomising principle as the modified burner (rotary cup, see Ch. 14). Its design virtually prevented recirculation of gases back to the flame front and thus reduced the effective residence time within the furnace. Recirculation was much more free with the modified burner. Drake and Hubbard[4,5] have examined the effect of swirl on particulate emission. With gas firing there are no particles to be emitted, the problem arises with firetube boilers in the presence of unburned gases at the entry to the convection tubes. Even small concentrations of carbon monoxide, from 500 p.p.m. upwards, can burn at the tube entries, enhance local heat transfer, and cause the tube ends to overheat and crack. For this reason, the carbon monoxide content of the gases entering the tubes should be not more than about 150 p.p.m. The precaution against problems of this kind is to restrict the volumetric heat release rate to about 1.9 MW m$^{-3}$ and to ensure that combustion is in fact complete within the furnace. Burners should be capable of producing compact flames and of recirculating gases in the furnace to secure good residence time.

### 4.1.5   Excess air

In the interests of fuel economy, excess air should be the minimum needed to ensure that complete combustion in the context of section 4.1.4 is achieved. Increasing excess air from 15% to 25% will decrease boiler efficiency by 0.6%, which, particularly in a large installation, is significant.

Having designed a burner which satisfies the preceding requirements, it could still be unsatisfactory in the matter of excess air. Figure 4.3 illustrates

| | |
|---|---|
| Fuel rich, long flame | |
| Air rich, short flame | |

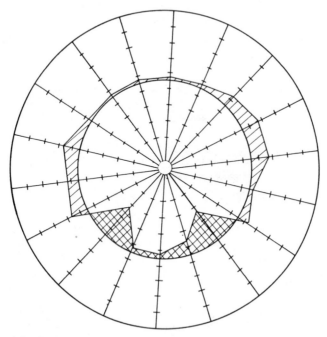

**4.3** Secondary air pattern in a dual fuel burner

the result of measuring air velocities at various sections in the throat of an oil burner. It has been assumed that the oil distribution is uniform, which it usually is within close limits. It will be seen that the air distribution varies over the circumference; in some regions the mixture is fuel rich and in others air rich. In the fuel-rich regions there will be incomplete combustion, which may or may not be removed by adequate recirculation. If it is not eradicated in this way the rich stream will emit smoke, which it is illegal to discharge from the chimney. To overcome this, sufficient additional air is provided to the burner to ensure that there is enough in the fuel-rich streams to ensure complete combustion. Some of this air will also enter the already air-rich streams where no more is needed. Hence, excess air is required to overcome the effects of a distorted air flow pattern.

Distortion of air flow in the burner can be caused by many factors, see section 14.3. The presence of sighting tubes, gas connections and pilot burners, within the burner can cause problems, but these are often of minor importance. The main disturbance often arises in the air supply to

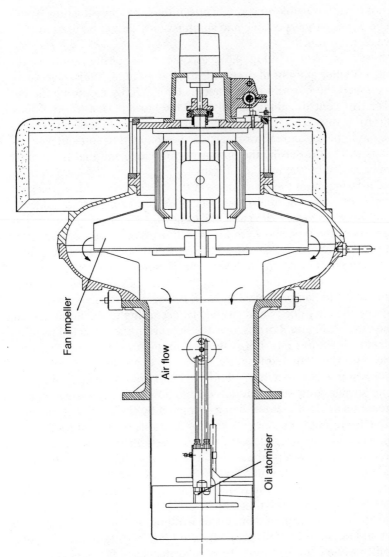

Fan impeller

Air flow

Oil atomiser

**4.4** Modern oil burner with axially mounted fan (From Dunphy Oil and Gas Burners Limited)

the burner: sharp changes of direction, unsuitably shaped or badly located dampers or a fan outlet too near the burner, all of which can cause marked distortion of the air discharge in the burner mouth. High air velocities amplify the distribution problem, so that, as a first step in correction, all ducts should be generously sized.

It is not always possible to avoid bends in the air supply ducting. These should be as smooth as possible and splitter vanes should be used to improve distribution.

The burner illustrated in Fig. 4.4 overcomes many of the problems, thus enabling a smaller amount of excess air to be used at all loads. The ducting which connects the fan with the burner is eliminated by mounting the fan, which is a true centrifugal fan, axially with the burner. Distortion of the air flow pattern is thus greatly reduced. Other advantages of this design are that the fan impeller and motor can easily be changed to suit site conditions without upsetting the burner performance and that the motor is effectively cooled by the inlet air stream.

In order to maintain the air velocity in the mixing zone at low loads, increased excess air is normally used to ensure good combustion under such conditions. Figures 4.5(a) and 4.5(b) illustrate an arrangement used with burners of the type shown in Fig. 4.4 to overcome this necessity. The area of the space between the diffuser plate (bluff body) and the air tube is reduced as the burner is turned down. This is accomplished by making that portion of the air tube in the locality of the bluff body of venturi shape and arranging that this can be moved forward to decrease the gap, thus maintaining the air velocity and vice versa, according to the input demand from the burner. Figure 4.5(a) shows the venturi in the retracted position equivalent to high fire; Fig. 4.5(b) shows it in the forward position for low fire. The venturi is actuated from the burner controls. Flow models, which use air or water as the visualising fluid can be of great assistance in designing burner equipment. Models enable flow patterns to be seen and measurements made, the latter being corrected by similarity criteria when scale models are used. For large boiler installations the use of flow modelling is now widely used, and enables excess air rates of 5% or less to be achieved in large water-tube boiler installations. Such low excess air rates cannot be used on firetube boilers, where the volumetric heat release rates are much higher, and the residence times less. Here excess air rates at high fire are normally in the range of 15–20%.

The shape and position of dampers controlling the air flow is important, these should be multi-bladed, with adjacent blades operating in opposite directions, i.e. be contrarotating. In circular ducts iris dampers are advantageous. Adjustable vanes may be used at fan inlets, but perhaps best of all is to use variable speed fans and to avoid the use of dampers except for closure of the boiler. The variable-geometry burner mentioned previously is also applicable.

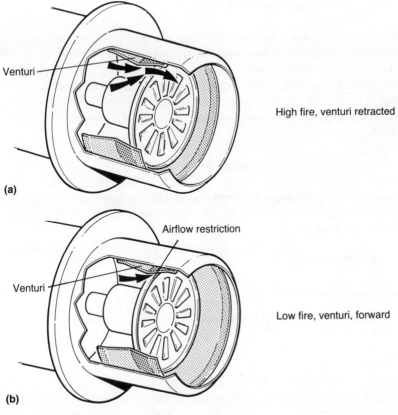

Venturi

High fire, venturi retracted

**(a)**

Airflow restriction

Venturi

Low fire, venturi, forward

**(b)**

**4.5**   Variable geometry burner
(From Dunphy Oil and Gas Burners Ltd)

### 4.1.6  Air preheat

If the temperature of a combustible mixture of air and fuel is raised, the burning velocity is increased. This assists in obtaining flame stability and reduced flame volume. Air preheat is often used with large water-tube boilers, the main purpose being to recover heat from the outgoing waste gases and to recirculate this back into the boiler system (section 7.4.2). Air preheat is not often used with firetube boilers, but it could be a field for future development.

Where air preheat is used there are two points that should be noted.

1. The amount of preheat should not be great enough to affect the velocity and fluid stability of oil, when this fuel is used.
2. Where sulphur-bearing fuels are used, air preheaters can be subject to corrosion by sulphuric acid (see Section 7.4.4).

## 4.2   The combustion of solid fuel

In industrial boilers, solid fuels, mostly coal, are frequently supported on
grates through which air passes, mixes with, and reacts with the fuel.
Alternatively, coal may be pulverised and, as with certain waste fuels, e.g.
sawdust, milled waste wood and palm fruit waste, may be burned in
suspension in air. The more recent development of fluidised bed
combustion is applicable to most fuels but, because of its novelty, caution
is advised.

### 4.2.1   The combustion of coal on grates

This is the commonest method of burning coal and has been in use for as
long as coal has been burned. There are three distinct modes of
combustion for coal.

#### 4.2.1.1   Overfeed combustion

This is the familiar mode in which fresh fuel is charged on to the top of that
which has already been ignited. It is illustrated in Fig. 4.6.

It will be seen that the freshly charged fuel is heated to ignition
temperature, partly by radiation from the ignited fuel and partly by
convection from the hot products of combustion arising from that fuel.
These products will be largely depleted of oxygen, which means that
volatile matter released by the fresh fuel cannot burn off. Overfire
(secondary) air must therefore be supplied to burn this, otherwise it will be
emitted as smoke (section 14.2.1.1). The heating of coal without much
combustion occurring at the same time enhances any coking tendencies the
coal may possess and this results in large lumps of coke which are difficult
to burn in the oxygen-depleted air. Overfeed combustion therefore tends
to produce smoke and coke, but ignition is rapid, and properly designed
appliances can be successful in boilers. Such combustion equipment is
described more fully in Chapter 14.

The ignition plane travels in the same direction as the air flow. Because
fuel is normally projected on to the fuel bed, fine particles will become
airborne and be carried through the system only partly burned. Overfeed
combustion is therefore more suitable for graded fuels where there will be
less carryover of partially burned fuel.

#### 4.2.1.2   Underfeed combustion

This is the reverse of overfeed combustion. Raw fuel is ignited from the
top purely by radiation, the ignition plane travelling downwards against the
air flow. Combustion of the fresh fuel is therefore not inhibited by lack of
oxygen, even though the ignition rate is slower, as it is not assisted by

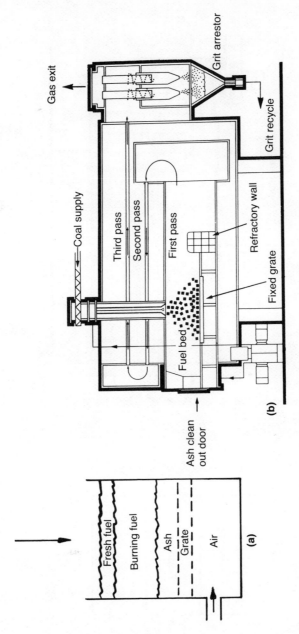

**4.6** Overfeed combustion: (*a*) principle, and (*b*) typical application (fixed grate)

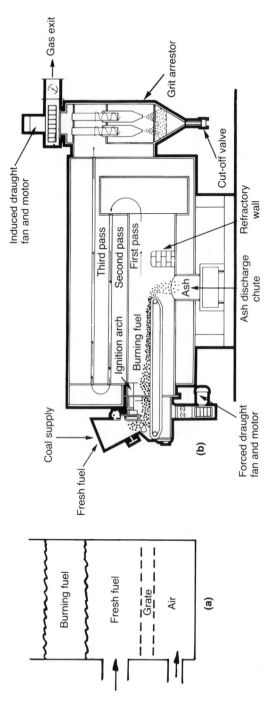

**4.7** Underfeed combustion: (*a*) principle, and (*b*) typical application (chain grate)

convected heat from fuel beneath. Smoke formation will be minimal and coke will be burned as it is formed, thus reducing the formation of large masses. Figure 4.7 illustrates the principle which is used in both firetube and water-tube boilers on travelling grates. Until the advent of pulverised fuel firing in the late 1940s, chain grates were the accepted method for firing power station boilers. Coal is fed gently on to the moving grate so that fine particles are not airborne, making this mode of combustion suitable for burning small coal. It is advantageous, indeed necessary, when firing small coal to provide sufficient surface moisture to bind the small particles to the larger pieces, otherwise the interstices between individual particles would be clogged and prevent the access of air, so interfering with the process of combustion.

The so-called 'underfeed stoker', more properly called the 'retort stoker' does not employ pure underfeed combustion. It really belongs to the next category but is rarely used on industrial boilers, its main application being to sectional boilers (which are not considered in this book).

### 4.2.1.3 Composite combustion

This combines underfeed and overfeed combustion by inserting fresh fuel into the heart of an already ignited fuel bed on a moving grate. It is illustrated in Fig. 4.8. The main advantages of this mode of combustion are that it ensures rapid ignition of the fuel, as in overfeed combustion, it can deal with small coal, but possibly not to the same extent as underfeed combustion alone, and it tends to form less smoke than does overfeed combustion. Against this is that, largely due to the action of the reciprocating grate used in practical appliances, coking tendencies are enhanced. This is advantageous when coals with little or no coking properties are burned, as the particles are bound together and thus stay on the grate instead of being blown off. The 'coking stoker' was invented in the 1870s specifically for this purpose. Modern versions of this appliance employ high burning rates and are used increasingly. There are two ignition planes, one overfeed and one underfeed, as illustrated in (Fig. 4.8(a). The relative characteristics of the three combustion modes are summarized in Table 4.1.

### 4.2.1.4 Grates

Grates in firetube boilers may be fixed or moving. Fixed grates need manual fire cleaning, while moving grates perform this operation automatically, although with certain coals some manual assistance may be needed on occasions. In both cases, however, grates support hot fires. Figure 4.9 illustrates temperature contours measured in an underfeed fuel bed. The grates themselves become hot, and adequate cooling must be provided, particularly when preheated air is used. The area of grate bars (or with chain grates, links) exposed to air flow must exceed by a large

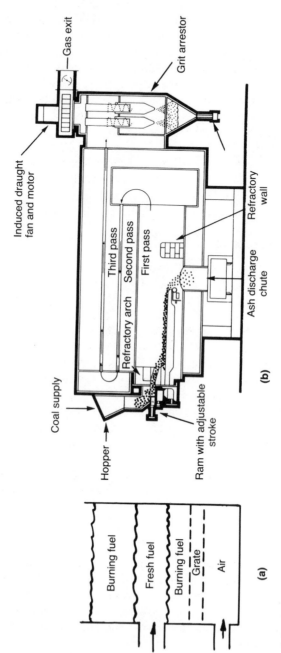

**4.8** Composite combustion: (*a*) principle, and (*b*) typical application (reciprocating grate)

**4.1**   Characteristics of overfeed, underfeed and composite combustion

| Mode of combustion | Advantage | Disadvantage |
|---|---|---|
| Overfeed | Rapid ignition, high response rate. | Smoke at low burning rates; not suitable for small coal. Can be sensitive to caking properties of coal. |
| Underfeed | Relatively smoke free, can burn strongly caking and small coals provided that these are properly conditioned with moisture and delivered uniformly over the grate. | Relatively slow ignition rate; can be difficult with coals with less than 20% volatile content. |
| Composite | Relatively smoke free, can burn small coal with up to 35% through 3 mm mesh. | Sensitive to caking properties of the coal. |

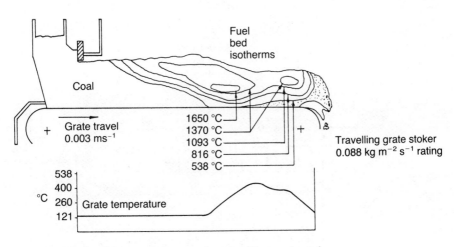

**4.9**   Fuel-bed temperature contours for travelling grate stoker
(After Marskell and Pratt *J.Inst. F.* 1952–53 **25**: 212)

margin the areas exposed to heat. For this reason the depth-to-width ratio of bars or links often exceeds 5 : 1; the cooling surfaces are often extended by incorporating nipples on them. More recently, cast iron, the basic material of construction, is being alloyed with chromium to provide greater resistance to oxidation at high temperatures. These alloy cast irons can however be brittle and must be formulated with care.

### 4.2.1.5 Air distribution through the fuel bed

As with the mixing of liquid or gaseous fuels with air, it is important to distribute the air in proportion to the needs of the fuel. This process is more difficult with the combustion of solid fuel on grates, and excess air rates are normally 40% or more. The cause of this is that with fixed grates, the fuel is never distributed with perfect uniformity, thin parts of the fuel bed pass more air than thick parts. Likewise, on moving grates, whilst the fuel may be evenly distributed across the grate at the beginning of its travel, the bed decreases in thickness and in fuel content towards the end and will there pass unreacted air. This problem is difficult to correct.

With travelling grates better control is achieved by arranging separate air supplies to different areas of the fuel bed. This has been done, over many years, with large travelling grates used in water-tube boilers, as illustrated in Fig. 4.10, but with the cylindrical furnaces of firetube boilers there is insufficient space beneath the grate for this to be done. A degree of control can, however, be achieved by restricting the air to the rear of the grate by a damper, as in Fig. 4.11(a), or by graded baffles, as illustrated by Fig. 4.11(b).

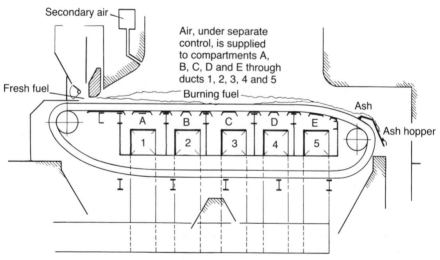

**4.10**   Travelling grate stoker for water-tube boiler

### 4.2.1.6 The effects of coal quality

Coal is a very variable substance, depending on its source of origin. Caking properties can be great or nil, and volatile content (which determines ease of ignition) may be as low as 7% (rarely used for industrial boilers) up to 40% with non-caking coals and 53% for lignite. Ash content may vary from less than 5% to over 40% with coal from some parts of the world,

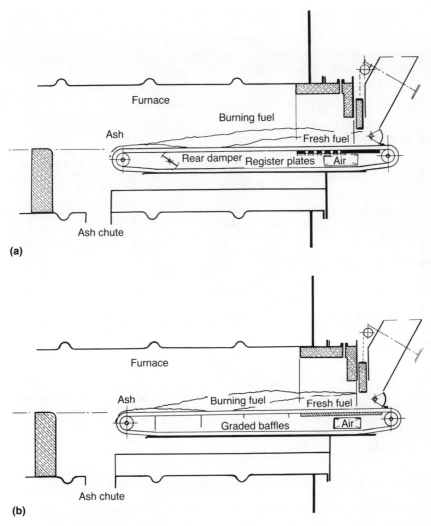

**4.11**  Chain grate for firetube boiler.
  (*a*)  grate showing dampers; and
  (*b*)  grate showing graded baffles

and particles of less than 3 mm in size (down to dust) may be less than 10% to over 50%. All these factors affect the combustion of the coal, the type of firing appliance used, and the resultant boiler performance. In the UK the coals preferred for industrial boilers are:

- 'singles' (25 mm × 12.5 mm) used mainly on overfeed, retort and coking stokers;
- 'washed smalls' (12.5 mm–0) used mainly on coking stokers and travelling grates; and

- 'untreated smalls' (12.5 mm–0) used mainly on large travelling grates or as pulverized fuel.

The caking and volatile properties are evaluated in the British Coal classification system shown in Table 4.2. In general, coals in the range of coal rank code 600–900 are preferred, coals with a lower code number being strongly caking or deficient in volatile matter.

**4.2** Coal classification (British coal grouping)

| Group | Types of coal | Caking properties | Coke type | Dry ash-free Volatile content per cent (%) |
|---|---|---|---|---|
| 900 | High-volatile | Non-caking | A–B | Greater than 30 |
| 800 | | Very weakly caking | C–D | Greater than 30 |
| 700 | | Weakly caking | E–G | Greater than 30 |
| 600 | | Medium caking | $G_1$–$G_4$ | Greater than 30 |
| 500 | | Strongly caking | $G_5$–$G_8$ | Greater than 30 |
| 400 | | Very strongly caking | Greater than $G_8$ | Greater than 30 |
| 301 | Medium-volatile | Very strongly caking | Greater than $G_8$ | 20–30 |
| 204 | Low-volatile | Strongly caking | $G_5$–$G_8$ | $17\frac{1}{2}$–20 |
| 203 | | Medium caking | $G_1$–$G_4$ | $15\frac{1}{2}$–$17\frac{1}{2}$ |
| 202 | | Weakly caking | C–G | 14–$15\frac{1}{2}$ |
| 201 | | Non-caking | A–B | $9\frac{1}{2}$–14 |
| 100 | Anthracite | Non-caking | A | Less than $9\frac{1}{2}$ |

Groups 900 down to 100 as shown in the above table form a consistent series with gradually changing caking properties and volatile content. Further information on grouping is given in Ref. 13.

Other characteristics which are of importance are as follows.

1. Ash fusion temperature should be as high as possible to avoid clinker formation (see Section 10.1.1).
2. Ash content does not necessarily affect boiler efficiency greatly, but high ash contents involve high transport and disposal costs.
3. Moisture content: coal burned on overfeed appliances should be low in moisture to avoid handling problems and to facilitate distribution on the grate; washed smalls for burning on travelling grates or coking stokers should have a relatively high moisture content, this being added when necessary to ensure mutual adhesion of coal particles. The ideal surface moisture for this category of coal is about 1.5% per 10% coal less than 3 mm.
4. If the contents of alkali metals in the ash substances are high, deposits are likely to form on the heated surfaces of the boiler (see Section 10.1.1)

In the 1950s, the British Coal Utilisation Research Association (now incorporated into British Coal) carried out an extensive investigation into the effects of coal characteristics on the performance of firetube boilers.[6,7,8] This work has been published and it is recommended that it be studied when firetube boilers are to be fired by coking stokers or chain grates. MacDonald[9] has discussed the performance of coking stokers in more detail. It is also recommended that the British Coal series of fourteen booklets 'Technical Data on Solid Fuel Plant' be studied.

### 4.2.2   Pulverised fuel (PF)

Most coal-fired power station boilers use pulverised coal, and many of the larger industrial water-tube boilers in industry also use this fuel. In the late 1940s attempts were made to use PF in firetube boilers. Some of these were partially successful in the older, more conservatively designed boilers but, later, decreasing size of furnace greatly restricted the heat-release rate obtainable from this form of combustion. However, in recent years there have been more optimistic developments in New Zealand where certain coals, because of their easy grindability and low ash content, are particularly suitable for this form of firing.[10] The combustion of PF has many similarities with that of fuel oil, although the fact that the particles are solid and have a lower volatile content makes their ignition and subsequent combustion slower processes. The mechanics of mixing are similar, although the transport of PF to the burner involves it being airborne. There is thus some degree of premixing.

Flame stabilisation can be carried out by the bluff-body device mentioned in section 4.1.2 of this chapter, but the geometry is slightly

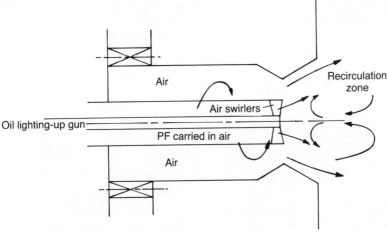

**4.12**   Principle of pulverised fuel burner

different to accommodate the considerably larger central fuel delivery duct needed. The system is illustrated diagrammatically in Fig. 4.12. Various other configurations are used, including multijet arrangements and axial swirl, which may be combined with the bluff-body principle.

### 4.2.3   Fluidised bed combustion (FBC)

This is a recent development for firing boilers and, although there are a number of installations using the principle, it is likely to be some time before it replaces conventional methods of firing.

   The principle involved is that a bed of inert, refractory sand-like material is caused to fluidise by passing air through it. The air may be heated by the burning of an auxiliary fuel, usually gas, before it enters the bed, or gas or oil may be burned above or within the bed. When the temperature of the bed has reached about 650 °C, coal is fed on to it or into it and ignites. The process now rapidly becomes self-sustaining, when the auxiliary fuel can be turned off. According to the demand on the boiler, the fuel and air inputs to the bed are increased until a maximum bed temperature of about 950 °C is reached. To avoid clinker formation and the emission of undesirable salts, the bed temperature should be controlled 950 °C or below; this is done by cooling the bed with water tubes which form an effective part of the heat-transfer surface of the boiler. Excess air is sometimes used for cooling, but this is undesirable as it increases the flue gas loss from the boiler. Recirculated flue gases may be used for this purpose, and avoid this problem, apart from minor losses from the ducting and fan casing. Figure 4.13 illustrates the principle of fluidised combustion, which holds promise of the following advantages.

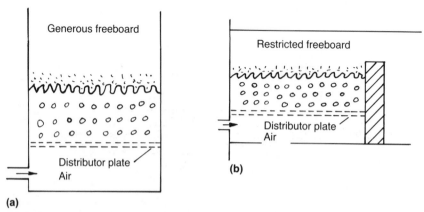

**4.13**   Principle of fluidised combustion
       (*a*) in a vertical firetube or water-tube boiler, and
       (*b*) in a horizontal firetube boiler

1.  It is a universal burner and can handle solid, liquid or gaseous fuels.
2.  Combustion can be controlled to 950 °C, which avoids the formation of clinker and emission of some undesirable substances.
3.  By incorporating dolomite or limestone in the bed material much of the sulphur in the fuel can be retained.

It is interesting to note that at any given time the amount of combustible material in the bed is only about 4% of the total mass. There are still problems to be solved: erosion of in-bed tubes and those surrounding the bed, and a high grit burden in the gases leaving the bed, which can cause erosion in subsequent passes. Fluidised bed combustion has been applied to both firetube and water-tube boilers, and to the composite design.

The water-tube construction has the advantage of greater flexibility in the design of the furnace shape, allowing greater freeboard in which entrained particles can drop back into the bed. It is still early in the development of FBC, but it seems likely that it will evolve, if slowly. Literature on the subject is extensive.[11,12]

# References

1.  Rose, J. W. and Cooper, J. R. *Technical Data on Fuel* (7th edn), 1977. British National Committee, World Energy Conference, London, pp. 253–66.
2.  Beér, J. M. and Chigier, N. A. *Combustion Aerodynamics*, 1972. Applied Science Publishers, London.
3.  BS 2790: 1986 *Shell Boilers of Welded Construction*. British Standards Institution, London. Also corresponding European (CECT) and International (ISO) standards now in draft.
4.  Drake, P. F. and Hubbard, E. H. 'Effect of swirl on the completeness of combustion', *J. Inst Fuel* 1963 (Sept.) **36**: 389.
5.  Drake, P. F. and Hubbard, E. H., 'Combustion system aerodynamics and their effect on the burning of heavy fuel oil', *J. Inst Fuel* 1966 (Mar.) **39**: 98.
6.  Gunn, D. C., 'The effect of coal characteristics on boiler performance', *J. Inst Fuel* 1952 (July) **25**: 148.
7.  MacDonald, E. J. and Murray, M. V. 'The effect of ash on the performance of shell boilers', *J. Inst Fuel* 1952 'A special study of ash and clinker'. **1**: 45.
8.  Murray, M. V. 'Performance of an overfeed coking stoker using $\frac{1}{2}''$–0 smalls' *J Inst Fuel* 1957 **30**: 276
9.  MacDonald, E. J. 'Smokeless combustion of 1/2 inch smalls on coking stokers', *Engineering and Boiler House Review* 1957 (Nov.).
10. Campbell, R. B. 'Pulverised fuel firing of shell boilers', *J Inst. E.* 1983 (June) **56**: p. 95.
11. Journals of the Institute of Energy for June 1979, **52** and December 1980, **53** are wholly relevant to this subject.
12. The Institute of Energy 3rd International Fluidised Bed Combustion Conference, London, 1984 (Oct.)
13. Rose J. W. and Cooper J. R. *Technical Data on Fuel* (7th edn), 1977. British National Committee, World energy conference, London. pp. 297–303.

# Heat transfer, fluid flow and water circulation

## 5.0  Introduction

The heat transfer to the water within a boiler and the circulation of that water around the boiler heating surfaces are interrelated; they will therefore both be dealt with in this chapter.

## 5.1  Heat transfer

The flow of energy resulting from a temperature difference has been the subject of research for many years. A detailed knowledge of the subject can be obtained by a study of the relevant literature. Well-known references in industry for the theory and application of the principles of heat transfer are McAdams[1] and Kern[2]. As will be seen from these and other publications, heat transfer is a complex and inexact subject, and there are varied opinions regarding which correlations to apply to a particular situation. Design engineers for heat-transfer equipment apply the basic principles but adapt them to suit their products, using results from tests carried out on test rigs or obtained from the operation of similar equipment over many years. This chapter can only deal with the subject briefly, but an attempt is made to give the reader an appreciation of the factors that influence the design and performance of boilers.

The transfer of heat occurs through the mechanisms of radiation, conduction and convection. In most practical cases it is by a combination of all these. Boilers, the purpose of which is to transfer the energy released by the combustion of a fuel to produce hot water or steam, involve all aspects of heat transfer.

In the case of heat transfer to the water, minimum resistance to heat flow is required to give minimum heated surface and hence cost, whereas to reduce the heat losses to the atmosphere through the enclosure, maximum resistance to heat flow is required.

## 5.2   Radiation

Heat transfer by radiation involves the transfer of energy through space in the form of electromagnetic waves, which differ from light in wavelength and frequency. Since radiant energy travels in straight lines with the velocity of light and may be absorbed, reflected or transmitted by a receiving surface in a manner similar to that of light, the laws of optics are important in the study of radiant energy transfer. The most well known example of this is the transfer of heat from the sun to the earth.

Radiant heat transmission can be luminous or non-luminous, that from a visible flame or visibly hot heat source such as red- or white-hot brickwork being known as luminous, that from a mass of hot gas in which there is no visible flame being non luminous. The only gaseous constituents of the products of combustion of conventional fuels giving off significant non-luminous radiation are carbon dioxide and water vapour, gases consisting of two or more different atoms.[3] More important are carbon and ash particles.[4] Heat transfer from the fuel burning in the combustion chamber of a hot-water or industrial steam generator is almost entirely by luminous radiation supplemented by a small proportion of convection.[4] The proportions depend upon the furnace design, the type of firing employed and the manner in which they affect the velocity with which the gas sweeps the heating surface.

One typical expression used to determine the heat transferred by radiation is developed from the Stefan–Boltzmann law[5], as follows.

$$Q = \sigma A E [(T_g)^4 - (T_w)^4] \qquad\qquad [5.1]$$

where:  $Q$   is the net heat transferred by radiation;
$\quad\quad\;\; \sigma$   is the Stefan–Boltzmann constant;
$\quad\quad\;\; A$   is the effective heat-emitting surface of the flame envelope and its enclosure;
$\quad\quad\;\; E$   is the resultant emissivity of the flame and absorbing surface;
$\quad\quad\;\; T_g$   is the flame temperature (absolute); and
$\quad\quad\;\; T_w$   is the cold surface temperature (absolute).

Equation [5.1] is based upon the theory that heat will be radiated from the surface of the flame envelope in proportion to the area and emissivity of that envelope and the fourth power of its absolute temperature, that some of the heat will be intercepted and re-radiated by the cold surface in proportion to its area, emissivity and the fourth power of its absolute temperature. The difference $Q$ represents the net heat interchange between the two. The effective surface area of the flame envelope will depend upon the fuel and type of firing equipment used, for example variations between boilers having burners in all four corners of the combustion chamber (tangentially fired) and those with burners in one wall only (wall-fired). Further variations will result over a range of heat inputs and with different furnace ratings, that is heat input per unit of volume or surface area.

The flame emissivity depends upon many factors among which are luminosity, non-luminous radiating constituents, volume and temperature.

The temperature of the flame envelope will not be constant or uniform in the water-cooled furnace of a boiler where a large proportion of the heat is absorbed as it is released, some 30–50% of the total input.[6] Combustion characteristics of the various types of firing equipment differ, some producing short, intense and highly turbulent flames while combustion with others is relatively slow, producing a comparatively long and lazy flame of greater volume. With oil and gas flames variations in air velocity and distribution can produce these effects. The effectiveness of water-cooling surface installed in the combustion chamber will depend upon the wall construction used[7] (see Fig. 5.1), the effectiveness of tube areas obscured by refractory materials and the firing appliance. With modern fully welded walls (Fig. 5.1(c)), the metal wall area exposed to the flame is considered to be 100% effective, as will be the exposed furnace tube area of a firetube boiler.

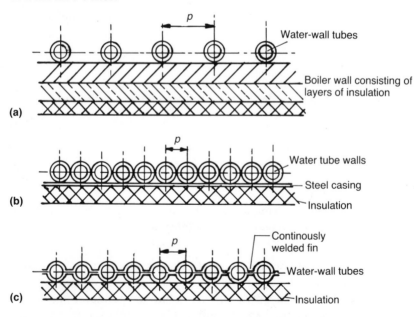

**5.1**   Typical method of constructing furnace water- walls.
   (*a*) Open-pitched tubes $p > (d + 2$ mm). Effectiveness is dependent upon tube diameter and pitch. (*b*) Tangent tube wall $p < (d + 2$ mm). 100% effective tube surface. (*c*) Welded tube wall. 100% effective tube surface

For any one furnace, the area of the 'cold' heat-absorbing surface is fixed, but its effectiveness can be altered by fouling and slag accumulation, therefore variation can occur over a period of operation which affects the performance of the whole boiler. Generally, the temperature of the clean, cold surface can be taken as that of the water inside the boiler without introducing an appreciable error. Slag accumulated will not only affect the heat-receiving area but also the temperature of the cold or heat-receiving surface.

The interest of the boiler design engineer in furnace performance prediction lies in obtaining an accurate assessment of the temperature of the gas leaving the furnace and also of the heat flux, i.e. heat transfer rate to the furnace walls. The temperature of the gas leaving the furnace is important to the design of the boiler convection heating surface, particularly that of the superheater when one is fitted. The heat flux to the walls is required for the mechanical design of the tubes and wall and also for the design of the boiler circulating system in the case of a water-tube boiler. When a superheater is installed in a position where it is exposed to furnace radiation, the radiant heat input has to be determined as accurately as possible to enable the heating surface required and maximum operating metal temperatures of the superheater to be calculated.

It will be apparent that because of all these factors the design engineer is not able to approach the problem in a purely theoretical manner with any degree of accuracy. The theory has to be supplemented with empirical data collected from units of similar design and operating conditions to obtain factors incorporating $\sigma$ and $E$ in equation [5.1].

This is achieved from furnace gas outlet temperature measurements which, in themselves, are difficult to determine due to gas flow and temperature maldistributions that exist. Local heat-flux measurements can also be taken, but provision has to be made in the boiler wall construction for insertion of probes and sealing when not in use.

To apply equation [5.1] to the calculation of the performance of the boiler, a heat balance for the furnace is carried out as follows:

net heat input = heat absorbed by radiation + heat in gases leaving furnace

$$H = \sigma AE(T_g^4 - T_w^4) + Mst_e \qquad\qquad [5.2]$$

Where  $H$  is the net heat input to the furnace;
   $M$  is the mass flow of gases;
   $s$  is the specific heat of the gases at temperature $t_e$; and
   $t_e$  is the temperature above datum of the gases leaving the furnace.

Other symbols are as defined for equation [5.1]. (All in self-consistent units.)
The solution of this expression is simplified by the use of graphical methods or by means of computer programs.

From equation [5.2] it will be apparent that increasing the heat absorbing surface $A$ of a furnace for a given heat input will increase the heat absorbed, and hence reduce the temperature of the gas leaving the furnace. Thus, by changing $A$ at the design stage, it is possible to arrive at an optimum value for $t_e$ at the inlet to the convection surfaces. This equation also takes into account the effective value of $E$ for the fuel which will be burned in the boiler. There can be considerable differences in this value, particularly between oil and gas. While increasing the area $A$ will increase the total heat transferred by radiation, $t_e$ will be lower, and hence the average heat flux will be reduced. For determination of the heat transferred by non-luminous radiation, equation [5.1] is applied using the relevant value for $E$, which depends upon the partial pressure of the radiating gaseous components and on the dimensions of the chamber in which the gases are enclosed.[8] Within a tube bundle, the tube diameter and configuration are controlling factors.

Obviously, as the gases pass through a boiler they are cooled and the value of $(T_g)^4$ falls, as will the amount of radiant heat transferred. In economisers and air heaters, it may be negligible compared with the heat transferred by convection.

## 5.3  Conduction

Heat is transferred by conduction through a material as a result of transfer of energy between molecules by molecular collisions. A high temperature signifies a high molecular energy which is transmitted to areas of low energy and hence low temperature. With opaque solids, heat transfer through them is by conduction only, heat transfer through liquids and gases will be by convection in addition as a result of natural convection currents within the fluid. The basic expression for heat transfer by conduction through a flat solid wall is developed from Fourier's law and with reference to Fig. 5.2 becomes

$$Q = \frac{k\,A}{x}(t_1 - t_2) \qquad\qquad [5.3]$$

Where: $Q$   is the quantity of heat transferred;

$\quad\quad\ \ A$   is the area of the heat transfer surface at right angles to the direction of heat flow through which the quantity of heat $Q$ flows;

$\quad\quad\ \ x$   is the thickness of the wall in the direction of heat flow;

$\quad\quad\ \ k$   is the thermal conductivity of the material forming the wall;

$\quad\quad\ \ t_1$   is the hot-face temperature of the wall; and

$\quad\quad\ \ t_2$   is the cold-face temperature of the wall.

(All in self-consistent units.)

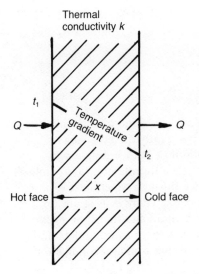

Thermal
conductivity $k$

$t_1$

$Q \rightarrow$

Temperature
gradient

$\rightarrow Q$

$t_2$

Hot face

$x$

Cold face

**5.2**  Conduction through a solid wall

Of the three modes of heat transfer, calculations for conduction tend to be
the simplest and most arithmetically accurate, since the only physical
property involved is the thermal conductivity of the material. This can be
determined for the various materials with a reasonable degree of accuracy.
Values of $k$ can be obtained from tables of physical properties of materials
in many readily available references.[9,10,11,12]

In a boiler, heat is transferred by conduction through refractory and
insulation used in the boiler enclosure, through the walls of the tubes
forming the heat-transfer surfaces and through any scale or deposits
formed on these surfaces. When considering conduction through the wall
of a cylinder the area through which the heat is being transferred is
continually changing as the radius of the cylinder changes. It can be shown
by integration that the expression for heat transfer becomes[13]

$$Q = \frac{k A (t_1 - t_2)}{\mathrm{Log}_e \dfrac{d_o}{d_i}} \qquad [5.4]$$

Where: $k$   is the thermal conductivity of the wall;
$A$   is the area of the heat receiving surface;
$t_1$   is the temperature at the inner face of the cylinder of inside
diameter $d_i$; and
$t_2$   is the temperature at the outer face of the cylinder of outside
diameter $d_o$.

This expression is used when considering the heat transfer through thick-
walled cylinders such as the insulation of pipework. For thin-walled

cylinders of the dimensions normally met in boiler tubes equation [5.3] can be used with an acceptable degree of accuracy.

## 5.4  Convection

Heat is transferred by convection between a surface and a fluid when the fluid is moving over the surface and being constantly mixed giving a relatively uniform temperature over the fluid flow path, i.e. flow is turbulent. There is, however, considered to be a thin, laminar (streamline) flowing film of fluid local to the static surface through which heat is transferred by conduction to the turbulent mass. There is a significant temperature gradient between the bulk of the fluid and the static surface, this is illustrated in Fig. 5.3. Convection heat transfer as we understand it therefore tends to be a combination of true convection and conduction.

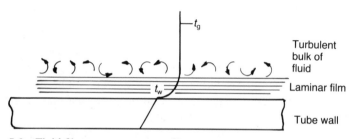

**5.3**   Fluid film temperature profile

Turbulent flow exists at Reynolds numbers above about 2000. Reynolds number is one of several dimensionless expressions used in fluid flow and heat transfer theory, see equation [5,7], which have contributed greatly to the development of knowledge on these subjects. Reynolds number is generally indicated by the symbol $Re$ which is evaluated by

$$Re = \frac{dG}{\mu} \qquad [5.5]$$

Where: $d$   is a characteristic dimension of the heat-transfer surface, for example tube diameter;

$G$   is the mass flow rate of the fluid past the heat-transfer surface (This is in mass units, per unit cross-sectional area through which the fluid flows per unit of time); and

$\mu$   is the dynamic viscosity of the fluid at its temperature in the locality of heat transfer.

In using these dimensionless relationships it is essential that self-consistent units are used, but providing that this is done, the value obtained is independent of the system of units, i.e. Reynolds number has the same value in SI units (which are used throughout this book) as in Imperial or any other units.

Since convective heat transmission always involves a flowing fluid, the laws of heat transfer by convection are closely related to the laws of fluid flow.[14] If the fluid flow is induced by a change of density of the fluid resulting from the transfer of heat to or from it, heat transfer is said to take place by 'natural convection'. A typical example of such a case is the transfer of heat to the atmosphere from a central heating radiator. If a fan or pump is used to create the fluid flow, heat transfer is said to take place by 'forced convection'.

With natural convection, which occurs at low fluid velocities and under conditions where the velocity profile is difficult to assess, heat transfer coefficients are difficult to predict theoretically. Fortunately for the boiler designers, in most applications heat transfer takes place by forced convection. At boiler design conditions the fluid is usually well mixed and the Reynolds number is above 2000. The basic expression, attributed to Newton, for heat transfer by convection from a hot moving turbulent fluid to a cool stationary wall is:

$$Q = h A (t_b - t_w) \qquad\qquad [5.6]$$

Where: $Q$   is the heat transferred;
      $A$   is the area of heat-transfer surface at right angles to the direction of heat flow through which the quantity of heat $Q$ flows;
      $t_b$   is the bulk temperature of the main stream of fluid;
      $t_w$   is the wall temperature; and
      $h$   is the coefficient of convective heat transfer, often called the 'film coefficient'.
     (All in self-consistent units.)

It will be seen that this is very similar to the well-known conduction equation

$$Q = \frac{k A}{x} (t_1 - t_2),$$

i.e. equation [5.3].

A common expression used to correlate the many variables that influence the heat transfer by forced convection for turbulent flow from a fluid to a tube wall and from which the heat-transfer coefficient is determined is written as follows

Nusselt Number $= a$(Reynolds Number)$^m$ (Prandtl Number)$^n$

$$\text{or} \quad \frac{h d}{k} = a \left( \frac{d G}{\mu} \right)^m \left( \frac{C_p \mu}{k} \right)^n \qquad\qquad [5.7]$$

Where: $h$   is the coefficient of convective heat transfer;
       $k$   is the thermal conductivity of the fluid.
       $C_p$  is the specific heat of the fluid.
       $a$, $m$ and $n$ are experimentally determined constants; and
       $d$, $G$ and $\mu$ are as defined in equation [5.5]
       (All in self-consistent units.)

Much work has been devoted to the selection of the values to be used for the constants $a$, $m$ and $n$ and also for the temperature at which the physical properties $k$, $\mu$ and $C_p$ should be evaluated[15], i.e. bulk fluid or film temperature.

Specific values for these constants and variations of equation [5.7] are proposed from a number of sources for individual flow patterns such as flow inside tubes, flow in annuli, axial flow outside tubes, flow externally across tubes and flow across flat plates. For the two most common examples in boiler work, flow through and flow at right angles across tubes, the following are examples.

Flow through tubes[16]

$$\left(\frac{hd}{k}\right) = 0.023\left(\frac{dG}{\mu}\right)^{0.8}\left(\frac{C_p\mu}{k}\right)^{0.4} \qquad [5.8]$$

Flow across staggered tubes[17]

$$\left(\frac{hd}{k}\right) = 0.33\left(\frac{dG}{\mu}\right)^{0.6}\left(\frac{C_p\mu}{k}\right)^{0.33} \qquad [5.9]$$

Equation [5.8] can be regarded as fundamental, but its practical employment involves a number of other factors, e.g. the added effect of gas radiation. Roderick, Murray and Wall[20] have described the results of extensive investigations of heat transfer in the tubes of firetube boilers and have presented the results in graphical form which can be very simply used for determining the heat transfer in such tubes. In Chapter 12 a numerical example of this method is given.

For flow across tube banks, equation [5.9], various values of $a$ and $m$ have been determined by Grimison.[18] Application of these expressions to experimental results indicates significant deviations[19], showing the need for boiler design engineers to rely on feedback of information from similar boilers to achieve the necessary degree of accuracy.

In a boiler, heat is transferred by convection from the hot gases flowing past the tube surface to the adjacent tube wall and from the tube surface to the water flowing past on the opposite side of the tube wall. The boiler design engineer has, through his choice of heating surface arrangement, some control over the gas mass velocity $G$ and the dimension $d$. All other factors being constant, the rate of heat transfer through the boundary layer, or the film coefficient, is essentially a function of the mass velocity, such that increasing the mass velocity by a factor of two will increase the film coefficient, resulting in a reduction of the heating surface required, for a given heat transfer, of about 20%.[20] It must be appreciated however that increasing the mass velocity by a factor of two will increase the pressure drop across the heating surface by a factor of four[20] and hence increase the power required to overcome this by a similar amount. Indiscriminate use of high gas velocities across tubes can introduce tube vibration problems.

By the application of equations [5.8] and [5.9] it can be shown that for a given gas velocity the heat-transfer coefficient increases as the tube diameter is decreased, e.g. flow through tubes from equation [5.8]

$$h \propto \left( \frac{1}{d^{0.2}} \right)$$

When $d = 76$ mm    $h \propto 0.42$
$\phantom{When\ }d = 50$ mm    $h \propto 0.46$

Flow across tubes from equation [5.9]

$$h \propto \left( \frac{1}{d^{0.4}} \right)$$

When $d = 76$ mm    $h \propto 0.177$
$\phantom{When\ }d = 50$ mm    $h \propto 0.21$

The choice of tube diameters and pitchings used in boiler work is restricted by various factors as follows.

### 5.4.1 Firetube boilers

The tube diameter should be large enough to avoid being blocked on the gas side by ash deposits. The number and diameter of the tubes should be selected to give a gas velocity below which tube erosion will not occur due to entrained dust in the gases. The minimum tube pitch that may be used is defined in BS 2790[21] and BS 855. These are based on the minimum amount of metal between tube seats in tube plates (ligaments) which is needed to accommodate the tubes and to allow their fixing by approved methods. These minima do not cause interference with water flow over the tube and hence do not cause heat transfer problems. Tube pitching affects the temperature of the metal in the tube seats: close pitching causing this to increase (see Appendix C of BS 2790).

### 5.4.2 Water-tube boilers

Boiler tubes should be of sufficient diameter to give adequate water circulation and mechanical stiffness. Superheater tubes should be of sufficient diameter to give acceptable steam side pressure drop, steam flow distribution and metal temperatures. Tube pitchings should be adequate to avoid buildup of deposits and bridging of gas passages.

There will be some gas mass velocity and tube configuration that will give an optimum arrangement, and it is the function of the design engineer to achieve the economical balance between cost of heating surface and cost of fan power within the limitations imposed by the fuel properties and the boiler configuration. Determination of the mean value of gas velocity is reasonably accurate when the gas flow is through tubes, as with firetube boilers. Many industrial water-tube boilers are of the multidrum type, of which Fig. 5.4 is typical. With these the gas paths may be formed by baffles placed in the tube bundles. Here the gas flow may not be at right angles to the tubes and there may also be a number of changes of direction where gas flow pattern is difficult to predict accurately. Equally difficult is the accurate prediction of the effectiveness of the heating surface.

The design of such boilers is, therefore, very dependent upon the feedback of performance data from other boilers to obtain suitable factors to be applied in the calculations for the achievement of acceptable results.

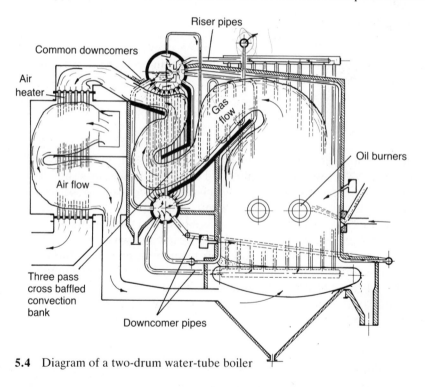

**5.4**   Diagram of a two-drum water-tube boiler

## 5.5   Application of heat-transfer formulae to boiler design

Let Fig. 5.5 represent a flat wall consisting of three distinct layers of solid material, through which heat is being transferred from a hot source of heat

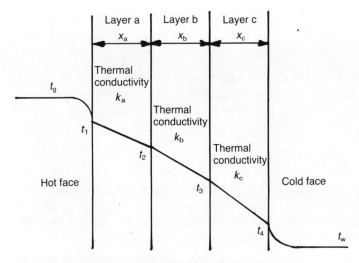

**5.5**  Temperature profile through a composite flat wall

at a temperature $t_g$ on the one side, to a cold fluid at a temperature $t_w$ on the other. This arrangement can represent a boiler wall consisting of three layers of insulating materials or it can represent a tube wall having a layer of deposit on the high-temperature gas side and a layer of scale on the low temperature water side.

The three layers a, b & c have thicknesses of $x_a$, $x_b$ and $x_c$, with thermal conductivities of $k_a$, $k_b$ and $k_c$. The face temperatures of the three layers are $t_1$, $t_2$, $t_3$ and $t_4$. The coefficient of heat transfer from the hot fluid to the hot face of the wall is $h_g$ and from the cold face of the wall to the cold fluid is $h_w$.

The heat flow $Q$ through a path of area $A$ across the wall can be represented by the following expressions:

$$Q = Ah_g(t_g - t_1)$$
$$= A\,h_a(t_1 - t_2) = A\frac{k_a}{x_a}(t_1 - t_2) \text{ from equation [5.3]}$$
$$= A\,h_b(t_2 - t_3) = A\frac{k_b}{x_b}(t_2 - t_3)$$
$$= A\,h_c(t_3 - t_4) = A\frac{k_c}{x_c}(t_3 - t_4)$$
$$= A\,h_w(t_4 - t_w)$$

When applied to boiler insulation these expressions can be used to determine the thickness of materials a, b and c required to give the desired heat loss $Q$ to atmosphere or the desired cold-face temperature $t_4$ on the outside of the boiler. Temperatures $t_1$, $t_2$ and $t_3$ should not exceed the maximum service temperatures of materials a, b and c, respectively. The heat transfer to the hot face in a boiler will usually be made up of components from radiation (luminous and non-luminous) and convection, hence

$$h_g = (h \text{ radiation} + h \text{ convection}) = (h_r + h_c).$$

To be able to sum these to give $h_g$ they must all be in the same units and relate to the same area and temperature differential. The coefficient $h_r$ for the radiation component can be determined by the following expression.

$$Q_{radiation} = h_r A(t_g - t_1)$$

$$\text{hence } h_r = \frac{Q}{A(t_g - t_1)} \qquad\qquad [5.10]$$

When carrying out heat-transfer calculations for heat exchangers and, particularly, boilers, it is more usual to use an overall coefficient of heat transfer which combines the effects of the deposits, scales and tube wall in a single term which reduces the expression for heat transfer to

$$Q = UA(t_g - t_w) \qquad$$ where $U$ is the overall coefficient of heat transfer.

$(t_g - t_w)$     = the sum of the temperature differences
the overall temperature     through the inside and outside films and the
difference.     layers of the wall.

From these can be developed the expression

$$\frac{Q}{UA} = (t_g - t_w) = \frac{Q}{Ah_g} + \frac{Qx_a}{Ak_a} + \frac{Qx_b}{A \, k_b} + \frac{Qx_c}{A \, k_c} + \frac{Q}{A \, h_w}$$

Multiplying through by $\frac{A}{Q}$ this becomes

$$\frac{1}{U} = \frac{1}{h_g} + \frac{x_a}{k_a} + \frac{x_b}{k_b} + \frac{x_c}{k_c} + \frac{1}{h_w} \qquad\qquad [5.11]$$

from which the overall coefficient of heat transfer $U$ can be determined.

Often the reciprocal of the heat-transfer coefficient of conduction $\frac{x}{k}$ is called the resistivity of the substance, and in the case of fouling and scales $\frac{x}{k}$ is referred to as the fouling resistance. Values of fouling resistance for various deposits are available[22,23], or as the coefficients $\frac{k}{x}$ .[24] It is still necessary to use the individual coefficients when determining the temperatures $t_2$ and $t_3$ necessary to enable tube materials and thicknesses to be selected in accordance with the boiler design codes.

All the examples so far referred to have assumed that the heat transfer is through a flat wall. Boiler heating surfaces are mainly cylindrical and hence the surface area through which a given quantity of heat $Q$ flows on the inside of a tube is less than that on the outside, see Fig. 5.6. Before

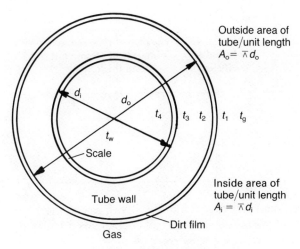

**5.6**  Section through tubular heating surface

calculating an overall coefficient from the individual coefficients in accordance with equation [5.11] it is necessary for these to be referred to the same surface area of the tube. This can be either the inside or the outside of the tube. (It is usually the surface swept by the gases.) Taking a water-tube boiler as an example, if the heat-transfer coefficient of the water film is $h_w$ on the inside of the tube, the transfer rate through the outer surface of greater area will be smaller for the same total heat flow rate, e.g.

$$Q = h_w A_i(t_4 - t_w) = h_{wo} A_o(t_4 - t_w)$$

where $h_{wo}$ is the coefficient referred to the outside of the tube

$$h_{wo} = h_w \frac{A_i}{A_o} = h_w \frac{d_i}{d_o}$$

$A_i$, $A_o$, $d_i$ and $d_o$ are defined on Fig. 5.6. For the heat-transfer coefficient through the tube wall, the mean diameter is used as the diameter to which the coefficient applies; hence for the overall heat-transfer coefficient through a tube wall referred to the outside tube surface, equation [5.11] becomes

$$U = \cfrac{1}{\dfrac{1}{h_g} + \dfrac{x_a}{k_a} + \dfrac{x_b}{k_b\left(\dfrac{d_i + d_o}{2d_o}\right)} + \dfrac{x_c}{k_c\left(\dfrac{d_i}{d_o}\right)} + \dfrac{1}{h_w\left(\dfrac{d_i}{d_o}\right)}} \qquad [5.12]$$

Typical values of heat-transfer coefficients as would apply to a boiler transferring heat from a gas through a tube to boiling water are given in Table 5.1. This table also shows the same coefficients referred to the

### 5.1  Typical Heat-transfer Coefficients

|  |  | Actual coefficient W m$^{-2}$ °C$^{-1}$ | Coefficient referred to outside diameter of 76 mm o.d. tube 4 mm thick W m$^{-2}$ °C$^{-1}$ |
|---|---|---|---|
| Gas film | $h_g$ | 80 | 80 |
| Gas-side fouling | $h_a$ | 6 000 | 6 000 |
| Tube wall | $h_b$ | 12 000 | 11 368 |
| Water-side scale | $h_c$ | 6 000 | 5 368 |
| Boiling water | $h_w$ | 12 000 | 10 737 |
| Overall | $U$ | 78.4 | 76.7 |

outside of a 76 mm outside diameter tube 4 mm thick with gas on the outside and boiling water on the inside. The overall coefficients calculated using equation [5.12] are also given.

From these values it can be seen that the accurately calculated overall coefficient of heat transfer is approximately 96% of the gas-side coefficient. It is useful to note that the overall coefficient of heat transfer is always less than the smaller of the individual coefficients. The whole procedure of determining the overall coefficient in such cases, where the gas-side coefficient is very small compared with the remainder, is simplified by applying a factor to the gas-side coefficient. This factor, determined by experience, is variously called 'fouling factor', 'surface factor' or whatever the design engineer wishes. It must be pointed out that the $h_c$ indicated for a water-side scale is a value that can be expected with correct water treatment. The effects of scale on the water side of boilers must not be underestimated. Figure 6.7 illustrates how the metal temperature of the combustion chamber of a firetube boiler increases with increasing scale thickness for a given heat flux. If flame impingement occurs in addition to the scale, giving a significant increase in the heat flux, the metal temperature will be increased further with possibly disastrous results.

In the case of superheaters and airheaters, where the fluid on both sides of the tube wall is a gas or vapour, the heat-transfer coefficients of both fluids are of sufficient magnitude to make it necessary for both to be taken into account in calculating the overall coefficient. The tube wall and fouling coefficients can be accounted for by the application of a suitable experimentally determined factor to the overall coefficient. The effect of two coefficients of similar magnitude is shown by the example of a tubular airheater having 60 mm outside diameter tubes 3 mm thick (54 mm inside diameter) with gas inside the tubes and air outside.

$h_g$ through the gas film    25 W m$^{-2}$ °C$^{-1}$
$h_w$ through the air film    45 W m$^{-2}$ °C$^{-1}$

The overall coefficient using equation [5.12], referred to the inside of the tube and neglecting fouling and tube wall coefficients is given by

$$U = \cfrac{1}{\left(\cfrac{1}{25}\right) + \left(\cfrac{60}{45 \times 54}\right)} = 15.45 \ \text{W m}^{-2} \ ^{\circ}\text{C}^{-1}$$

## 5.6   Temperature difference

So far, all examples have referred to a specific point in a boiler with a constant temperature difference. In practice, there is a continually changing temperature on the gas side of a boiler as heat is transferred and the gas is cooled. There is also a changing temperature on the cold fluid side where there is no change of phase, i.e. no change from water to steam, as in economisers, superheaters and airheaters. In evaporators, where steam is being generated the water side temperature is constant.

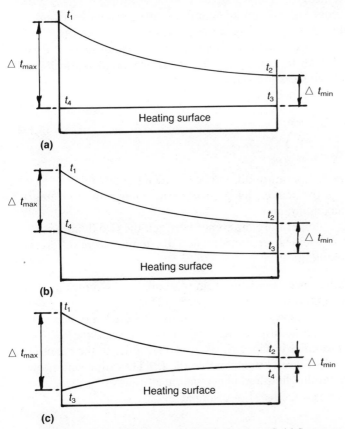

**5.7**   Temperature difference profiles for various fluid flow patterns.
(*a*) Evaporation of liquid at constant temperature. (*b*) Counter flow.
(*c*) Cocurrent (parallel) flow.

The three possible cases that occur within a boiler are illustrated in Fig. 5.7. Figure 5.7(*a*) illustrates the temperature profiles of the fluids in a heat exchanger in which evaporation of a liquid is occurring at constant pressure, and hence temperature, by heat transfer from a hot gas or liquid that is flowing across the heat transfer surface. A common example would be the generation of steam in a boiler as the hot products of combustion flow across the tubes (note there will be a small change of temperature due to pressure variations caused by static head and fluid pressure drop). Figure 5.7(*b*) illustrates the temperature profiles in a heat exchanger in which the fluids are flowing in opposite directions through a heat exchanger without change of phase. This is known as a counterflow exchanger. Figure 5.7(*c*) illustrates the temperature profile in a heat exchanger in which the fluids are flowing in the same direction through the exchanger without change of phase. This is known as a parallel or cocurrent flow exchanger. This arrangement can be used in superheaters or airheaters when metal temperatures are critical.

As can be seen from Fig. 5.7, the temperature difference varies from inlet to outlet of the exchanger giving a varying rate of heat transfer. This temperature difference variation is not linear: the average difference for the purposes of heat transfer calculations can be shown mathematically to be given by the following expression.[25]

Mean temperature difference

$$\Delta t_{mean} = \frac{\Delta t_{max} - \Delta t_{min}}{\text{Log}_e \left( \dfrac{\Delta t_{max}}{\Delta t_{min}} \right)} \qquad [5.13]$$

Where: $\Delta t_{max}$ is the temperature difference between the two fluids at that end of the heat exchanger where the temperature difference is maximum; and

$\Delta t_{min}$ is the temperature difference between the two fluids at that end of the heat exchanger where the temperature difference is minimum.

The temperature difference calculated from this equation is known as the 'logarithmic mean temperature difference' or LMTD.

Figure 5.8 illustrates a typical temperature profile through a boiler having a superheater, economiser and an airheater to indicate the application of the examples in Fig. 5.7. By applying values to the examples in Fig. 5.7(*b*) and 5.7(*c*) it can be shown that the LMTD with counterflow is higher than with parallel flow and hence that heat transfer by counterflow is the most effective. For example let

$t_1 = 1132 \text{ °C} \qquad t_2 = 724 \text{ °C}$
$t_3 = 280 \text{ °C} \qquad t_4 = 490 \text{ °C}$

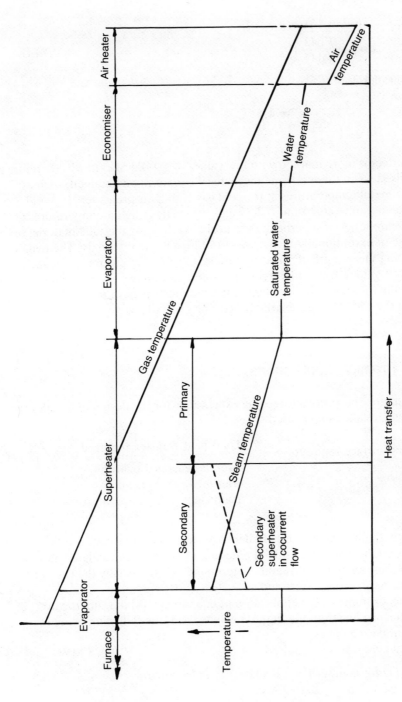

Furnace Evaporator | Superheater | Evaporator | Economiser | Air heater

Temperature

Gas temperature

Secondary

Primary

Secondary superheater in cocurrent flow

Steam temperature

Saturated water temperature

Water temperature

Air temperature

Heat transfer ⟶

**5.8** Temperature profiles through the convection surfaces of a typical water-tube boiler

For counterflow

$$\text{LMTD} = \frac{642 - 444}{\text{Log}_e \left(\frac{642}{444}\right)} = 537\ °\text{C}$$

For parallel flow

$$\text{LMTD} = \frac{852 - 234}{\text{Log}_e \left(\frac{852}{234}\right)} = 478\ °\text{C}$$

This effect is more marked with low values of $\triangle t$. Parallel flow is therefore only used when there is some mechanical benefit to be obtained, such as higher metal temperatures at the cold gas end of air preheaters to avoid low-temperature corrosion (Ch. 7 and 10) or lower metal temperatures at the steam outlet of superheaters to avoid overheating the tube material or the use of exceptionally thick tubes or of high grade materials. The penalty for doing this is a greater heating surface to achieve a given total heat transfer. Where true countercurrent or cocurrent flows are not achieved, as is usually the case in boiler plant, correction factors have to be applied to adjust the LMTD[26]; these are also given by McAdams.[27]

## 5.7 Practical applications

To combine the various foregoing expressions for the design of a boiler the following relationship is used

$$\text{Heating surface required} = \frac{\text{Total heat transfer required}}{\substack{\text{overall heat-} \\ \text{transfer} \\ \text{coefficient}} \times \substack{\text{Logarithmic} \\ \text{mean} \\ \text{temperature} \\ \text{difference}}} \qquad [5.14]$$

When designing sections of a boiler to perform a desired duty all the factors on the right-hand side of equation [5.14] can be calculated and the heating surface required readily determined. This is usually the case with superheaters, economisers and airheaters.

As an example of the application of equation [5.14] the superheater performance illustrated in Fig. 5.8 will be used, the following being the relevant design data.

Total heat transfer required          19.285 MW

LMTD for counterflow calculated as an example in section 5.6          537 °C

Overall heat-transfer coefficient (clean), section
5.9                                                   $87.3 \text{ W m}^{-2} \text{ }^\circ\text{C}^{-1}$

Fouling factor                                        0.85

From equation [5.14] the heating surface required

$$= \frac{19\ 285\ 000}{87.3 \times 0.85 \times 537} = 484 \text{ m}^2$$

The determination of the coefficient of heat transfer involves the use of gas
and steam quantities and the boiler dimensions, the selection of tube
diameter, spacings and number of tube entries for steam flow; the last to
ensure that the steam pressure drop across the superheater is acceptable.
The selection of these is explained in more detail in Chapter 13.

With standard boilers, such as firetube and shop-assembled water-tube
types where the boiler dimensions and hence convection heating surfaces
are fixed by the range of boilers for which drawings are available, the heat
transferred through that fixed heating surface has to be determined for the
boiler design conditions in order to obtain the gas temperature for any
heat-recovery equipment to be included.

The temperature of gas leaving that boiler heated surface is unknown,
and hence so are the heat transfer and logarithmic mean temperature
difference. Under such circumstances, equation [5.14] has to be solved by
trial and error by selecting values for the gas outlet temperature and
correcting the heat transfer until a balance is obtained. With the use of
computers to perform boiler design calculations this can be readily
achieved.

A similar procedure has to be used for all sections of the boiler where it
is required to determine the performance at an output other than the one
for which the design calculations were carried out. Experience enables a
design engineer to make a close first approximation, thus simplifying the
procedure significantly since it becomes unnecessary to correct the heat-
transfer coefficient for the variations of temperature of the gases.

## 5.8   Extended surfaces

In those parts of a boiler where heat transfer from the gases to water is
largely by convection, the gas-side heat-transfer coefficient is low
compared with those through the metal and on the water side, and
becomes the controlling factor in the magnitude of the overall coefficient.
If this can be increased in some way, the space occupied by the heated
surface for a given heat transfer can be significantly reduced. Alternatively,
more heat can be transferred in the same space. A common method of

achieving this is to use extended surfaces on the gas side, consisting of various forms of fins or gills attached to the tubes which increase the heated surface considerably per unit length of tubing. Typical examples of such surfaces as applied to economisers are illustrated by Fig. 7.11.

Air heaters which have a gas on both sides of the heat-transfer surface can benefit by extending the surface on both sides. Internally and externally finned cast-iron tubes are available, as are plate type heaters with extended surfaces on both sides of the plates. The additional surface does not increase the heat transfer proportionally due to the fact that the fin temperature rises as its distance from the parent tube increases, thus reducing the effective temperature difference. A 'fin efficiency' is introduced to correct for this.[28]

If extended-surface tubes are used in high gas temperature zones, e.g. furnaces using welded walls, care has to be taken to ensure that excessive metal temperatures do not exist. Methods of calculating heat transfer with extended surfaces are given by McAdams[20] and Kern.[30] The manufacturers of the various forms of extended surface tubes produce correlations applicable to their own fin configurations.[31,32] The manufacturers' data should be used where possible, since the data will have been determined from tests using tube pitchings and velocities as near as possible to those being applied in practice.

Extended surface tubing cannot be applied indiscriminately because tubes with closely pitched fins, say of 4 mm pitch, will be prone to fouling on the gas side of solid-fuel fired boilers and in the low temperature areas of oil-fired boilers, e.g. economisers. It is recommended that where extended surface tubes of the type illustrated in Fig. 7.11 are used with dust-laden gases the gas flow be vertical across horizontal tubes in order to avoid dust settlement on the fin surface, as would occur with vertical tubes and horizontal fins. It is for this reason that extended surface tubing is rarely, if ever, used on the evaporator and superheater surfaces of fossil-fuel fired boilers, particularly those of the multi-drum type using vertical or near vertical tubes.

Extended surfaces are used extensively on waste-heat boilers handling clean gases, particularly those associated with the heat recovery from the exhaust gases from gas turbines, see Chapter 16. Where the gas flow is through tubes, as in firetube boilers, an important parameter is length-to-diameter ratio; the greater this is, the greater is the temperature drop through the tube. Various devices have been used to increase the effect of length, tubes being contoured in sinuous or twisted fashion to give increased heat-absorption area and a greater effective length of gas travel. Alternatively, gases can be swirled by turbulators otherwise known as retarders. These cause the gases to cover a greater length of flow at increased velocity than in straight flow. Evans and Sarjant[33] carried out experiments which enabled a theory to be derived for estimating the effect of twisted steel tapes inserted into the tubes. Spalding[34] more recently evaluated the effects of different types of turbulators in comparison with plain tubes. The short answer to whether or not to use such devices is that it is always better, in terms of the increased draught loss which these devices incur, to use tubes of smaller diameter. This can be done at the design stage, but the use of turbulators may be justified when it is desired to improve the efficiency of existing boilers. Any kind of device to enhance heat transfer is, however, both an expense and a complication, as it will need to be removed for tube cleaning and afterwards replaced. Such devices are, however, used in certain types of hot-water boilers, where, to economise in space, short boilers are often required.

In highly rated areas of the combustion-chamber tubes of water-tube boilers, rifled bores have been applied to improve the internal water-side coefficient of heat transfer and hence to reduce the metal temperature below what it would have been with plain tubes. The heat transfer in firetube waste-heat boilers that have over-performed has been reduced by the introduction of tubular sleeves inside the boiler tubes which leave a small space between the two to give a static layer of gas acting as an insulator.


## 5.9  Practical examples

The determination of metal temperatures of the various heated surfaces is carried out by the application of equations [5.3] [5.6] and [5.12]. A typical example follows

Consider a superheater tube for which the relevant technical data is as follows.

| | |
|---|---|
| Tube outside diameter | 52 mm |
| Tube thickness | 5.89 mm |
| Gas-side coefficient of heat transfer | 95 W m$^{-2}$ °C$^{-1}$ |
| Steam-side coefficient of heat transfer based on   inside diameter of tube | 1820 W m$^{-2}$ °C$^{-1}$ |
| Gas inlet temperature | 1132 °C |
| Steam outlet temperature | 490 °C |
| Flow pattern | Counterflow |
| Tube material 1% chromium, thermal   conductivity | 30.6 W m$^{-2}$ °C$^{-1}$ |

The overall coefficient of heat transfer $U$ for clean conditions is

$$U = \cfrac{1}{\cfrac{1}{95} + \cfrac{1 \times 52}{1820 \times 40.22} + \cfrac{5.89 \times 52}{1000 \times 30.6 \times 46.11}}$$

$$= \frac{1}{0.011\,45} = 87.3 \text{ W m}^{-2}\,^{\circ}\text{C}^{-1}$$

Heat flux at the gas inlet/steam outlet

$$= U \times t = 87.3\,(1132 - 490) = 56\,047 \text{ W m}^{-2}$$

Temperature drop through the steam film

$$= \frac{\text{Heat flux}}{\text{Steam coefficient}} = \frac{56\,047}{1820} = 31\,^{\circ}\text{C}$$

Temperature drop through tube wall

$$= \frac{\text{Heat flux}}{\text{Tube-wall coefficient}} = \frac{56\,047}{4608} = 12\,^{\circ}\text{C}$$

Maximum tube metal temperature

$$= \text{Final steam temperature} + \text{Steam film } \triangle t + \text{tube wall } \triangle t$$
$$= 490 + 31 + 12 = 533\,^{\circ}\text{C}$$

This excludes any allowance that the design engineer may add as required by the design code[35], to allow for the steam temperature in individual tubes being above average due to gas flow and temperature and steam flow gradients across the boiler. It is also normal for the steam temperature to be subject to a tolerance of say $\pm$ 10 °C; this has also to be added to the final steam temperature used. All the above can give a tube wall design temperature somewhat above the 533 °C calculated above. The maximum allowable design stresses at various operating temperatures for the different materials are given by the relevant boiler design codes.[21,35]

## 5.10   Pressure drop in boiler flow systems

To create the flow of a fluid through or across the tubes of a boiler necessitates a differential pressure or a pressure drop from the inlet to the outlet of the flow path. This pressure drop usually has to be overcome by the use of pumps or fans which consume power in the process. Pressure drops are therefore an important factor in the design of a boiler and have to be a minimum consistent with good design and optimum cost. High pressure drops are a consequence of high velocities. As already indicated (section 5.4), a high pressure drop and hence high gas velocity usually give the minimum heating surface but maximum power consumption.

The boiler cost is a once and for all payment and is predictable, but power continues to be used throughout the life of the plant and hence increased power consumption can result in a considerable long-term cost. It is the duty of a boiler designer to make an evaluation of this based upon current information given by the customer. If the plant has a relatively low load factor, such as a seasonal space-heating system, higher pumping and fan costs may be tolerated than would otherwise be the case for a plant operating continuously throughout the year.

### 5.10.1   Flow through a pipe or tube

Frictional pressure drop resulting from the flow of a fluid through a pipe or tube is determined using an expression derived from the Fanning equation,[36,37] an example being as follows.

$$\text{Frictional pressure drop} = \frac{4fG^2 v\, L}{2gd} \qquad [5.15]$$

Where: $f$  is the friction factor;
  $G$  is the mass velocity (mass per unit cross-sectional area per unit time);
  $v$  is the specific volume of the fluid;
  $g$  is the gravitational acceleration;
  $L$  is the length of the tube; and
  $d$  is the internal diameter of the tube.
  (All the above in self-consistent units.)

This expression is for an incompressible fluid, additional terms are added to allow for changes in specific volume through the tube, such as can occur with large values of pressure drop relative to the actual pressure of the fluid.[38] Changes in specific volume also occur in superheater tubes as the temperature of the steam increases, and in evaporator tubes where a change of phase from water to steam is taking place. The friction factor $f$ is a function of Reynolds number and can be obtained from the well-established curve presented by Moody.[39] The friction factor given by Moody is for use in the Darcy formula

$$\text{Pressure loss} = \frac{fL}{d} \cdot \frac{V^2}{2g}$$

and is four times the value given by McAdams[40] for use in the Fanning equation. In this instance $V$ is the linear velocity. Care must therefore be taken in the correct application of friction factor to pressure drop calculations.

In calculating the overall pressure drop of a system, allowance has to be made for losses through bends, expansions, contractions and other fittings which can contribute a significant proportion of the pressure drop of a boiler circuit. This is particularly the case in superheaters and economisers where the tubes are arranged in banks and incorporate a large number of bends. In the case of pipework, the valves and other fittings also have to be allowed for. This can be done by increasing the effective length of the circuit used in equation [5.15] by an equivalent length for each item. These equivalent lengths are determined experimentally for each fitting and are usually expressed in the form $L/d$, where $d$ is the inside diameter of the pipe, tube or fitting. Values of these are available from a number of sources.[41,42]

Depending upon the complexity of the system being considered, the total pressure drop may consist of the sum of a number of separate expressions determined from equation [5.15] to enable the correct values of $f$, $G$ and $v$ to be applied to each section of the circuit should there be any variations in diameter and specific volume along its length. A typical frictional pressure drop could therefore be

$$\Delta p = \frac{4f_1 G_1^2 v_1 L_1}{2gd_1} + \frac{4f_2 G_2^2 v_2 L_2}{2gd_2} + \ldots$$

### 5.10.2   Flow across tube bundles

The pressure drop (or draught loss) resulting from a fluid flowing across tubes is more difficult to predict than is flow through tubes, there being a continuously changing velocity as the various rows of tubes are crossed. It is also influenced by the tube spacing and configuration and the type of tubing used, either plain or extended surface. As indicated in section 5.4, in water-tube boilers, the gas flow across the tubes is frequently not at right angles to the axis of the tubes and there may be changes of direction of flow within the tube bank, see Fig. 5.4. This all tends to make accurate pressure drop determination difficult.

An expression for the friction pressure drop with gas flow across and normal to the axis of tubes is

$$\Delta p = \frac{4fG^2 vN}{2g} \qquad\qquad \text{[5.16]}^{43}$$

Where: $N$  is the number of rows of tubes being considered;
      $G$  is the maximum gas mass velocity determined at the minimum flow area;
      $f$  is the friction factor for which values are given by Grimison for various in-line and staggered tube arrangements with various diameters; and
      $v$  and $g$ are as for equation [5.15].

It will be seen that there is a strong similarity between equations [5.15] and [5.16], in both of which the pressure drop is proportional to $G^2$. Much work has been done on this subject.[44] Manufacturers develop their own correlations for the tube configurations commonly used by them, both from experimental work and from practical experience. This is particularly the case with the various configurations of extended-surface tubing available.

## 5.11  Boiler water circulation

### 5.11.1  Natural circulation

Most industrial steam generators, be they firetube or water-tube. rely upon natural circulation of the water within the boiler to convey the heat transferred through the tube wall away from the water-side surface of the tube, thereby keeping the metal temperature at an acceptable level. The principle of natural circulation can be demonstrated in simple terms by considering a single tube of a water-tube boiler circuit, as illustrated in Fig. 5.9.

If the system is full of water at saturation temperature and heat is applied to the right-hand leg A or riser, the water will boil as the heat is absorbed in the form of latent heat, thus changing some of the water to steam. This steam will exist in the form of bubbles in the water, and this mixture is known as two-phase flow. Due to the low density (high specific volume) of steam relative to water (Fig. 6.2), the pressure at the datum point B created by the head $H$ of the steam and water mixture will be less than that due to the column of saturated water in the left-hand leg C. This situation cannot, of course, exist and a flow is created in the upward direction in the riser as indicated. This is of such a magnitude that the total pressure drop or resistance to flow from the steam drum, through the downcomer and up through the riser back to the steam drum, will be equal to the difference in head between columns A and C in the equilibrium condition.

The quantity of water circulating around the system must be several times the total quantity of steam generated in the heated riser to ensure that acceptable conditions exist. These will vary from about 4 times the evaporation at very high pressures (200 bar) up to as high as 30 times at low pressures. This number is known as the circulating factor of the circuit and is equal to

Total water mass flow into the tube
———————————————————
Mass of steam generated in that tube

The heat transfer process from the tubes to the water can be quite complex.

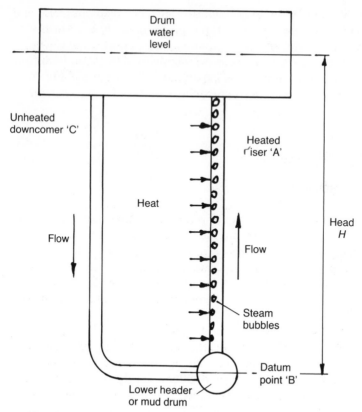

**5.9** Simple natural-circulation boiler circuit

Boilers are normally supplied with feedwater at a temperature below saturation temperature at the boiler working pressure. This, when mixed with the saturated water in the boiler, will result in the water in the downcomer being below the saturation temperature, or subcooled; the amount will depend upon the feedwater temperature entering the drum and the rate of water circulation around the boiler circuit. As a consequence of this and the higher static pressure at the bottom of the riser (point B of Fig. 5.9) than that existing in the steam drum, due to the static head of the water in the downcomer, there will require to be a certain amount of heat transferred to the water to raise it to saturation temperature. While heating the subcooled water to saturation temperature the heat transfer to the main body of water adjacent to the tube surface may be by convection or by the generation of steam bubbles on the tube surface which then pass into the body of the water by turbulence and condense, giving up their latent heat as sensible heat to the water.

It is particularly important in the design of hot-water boilers, especially when the static head is high, that the temperature of the water in contact with the metal surface is taken as the saturation temperature at the imposed pressure. In regions of high heat flux, convection alone cannot carry away the heat and this has to be done by local boiling. Failure to observe this can result in serious underestimate of metal temperatures, particularly in the case of firetube boilers where metal thicknesses are greater than in water-tube boilers.

When the water is all at saturation temperature, nucleate boiling takes place when the quickly formed steam bubbles break away from the tube surface and allow more water to reach the tube and hence maintain a high heat-transfer coefficient and low metal temperature. At high heat fluxes the bubbles do not readily break away from the tube surface and merge to form a film. This results in a reduced heat-transfer rate between the tube and water (water-side) and high metal temperatures. This is known as film boiling. The demarcation between the two processes is known as the point of departure from nucleate boiling (DNB). For industrial water-tube boilers other than of the once-through type, operation should always be made to be in the nucleate boiling regime by careful design of the circulation system.

There are a number of criteria used for establishing a high water-side heat-transfer coefficient and hence the safe water flow through the boiler circuits; a considerable amount of research has been carried out to determine these safety critieria. Some factors considered are: the water inlet velocity in the tubes, the mixture mass velocity and the percentage of steam by volume at the exit of the heated tube.

While the principle of circulation calculations is simply one of hydraulics, the actual calculations can be quite complex due to the fact that the heat input over the length of a heated tube is rarely at a uniform rate, thus making the determination of static head in the riser difficult. An added complication in most natural-circulation water-tube boilers is that some parts of the system are common to all circuits. For example, referring to Fig. 5.4, it will be seen that the convection bank tubes and all combustion chamber tubes (on the front, side and rear walls) are supplied with water from the lower or mud drum. The water from the steam drum all flows to the mud drum through a row of interdrum downcomers which are therefore common to all circuits. A further factor to be taken into consideration is that in vertical or near vertical tubes, the steam bubbles formed travel up the tubes faster than the water, due to the difference of density between the steam and water. The difference between steam and water velocities is greater at low pressures where the density difference is greater. This phenomenon is known as 'slip' or 'separated flow', compared with homogeneous flow where the steam and water are assumed to flow at the same speed.

With reference to Fig. 5.4, it will be seen that the heated tubes in the furnace are fed by external downcomer or feeder pipes and the steam and water mixture from the side wall circuits is returned to the steam drum by riser pipes. There is naturally a desire to reduce the number and size of these pipes to a minimum for minimum cost. The size and number can be optimised by study of the circulation characteristics of the circuits concerned. Good design points for natural-circulation water-tube boilers are:

1. minimise bends in the tubes;
2. avoid tubes having low heat input at the inlet end and high heat input at the outlet; and
3. avoid large differences in the water-side resistance of connected parallel circuits, since this will result in a high water flow in the tubes having a low pressure drop to achieve the desired flow in the high pressure drop ones, thus necessitating a larger number of risers and downcomers than would be required for separate circuits.

Firetube boilers generating steam or hot water operate under natural circulation, but the pattern of circulation is difficult to calculate due to the variation of heat input along each pass of tubes and also in the tube layout. Experience plays a large part in determining suitable tube layouts for such boilers.

### 5.11.2  Forced and assisted circulation

When very high steam pressures are used, 140 bar and upwards, and on special boilers where the tube circuits are such that the water flow is not continuously upwards and hence is not suitable for natural circulation, a pump is introduced into the boiler circulating system to generate the head necessary to create the desired water flow. Such systems operate under 'forced' or 'assisted' circulation. Figure 2.6 illustrates a typical forced-circulation boiler design. These are not to be confused with 'once-through' boilers (Fig. 2.7) where the feedwater pumps create the necessary head to overcome the boiler circuit pressure drop.

Forced or assisted circulation gives the freedom to use fewer tubes of smaller diameter with more complex circuitry within the boiler, the latter being particularly useful with waste-heat boilers (Ch. 16). To give the correct water flow distribution between tubes operating in parallel, it is usual for nozzles, orifice plates or restrictors to be placed in the pipework for each circuit; this is illustrated diagrammatically by Fig. 5.10, the size of the orifice being selected to give the desired pressure drop and hence water flow to the various circuits. La Mont boilers have nozzles at the inlet of each boiler tube.

The pressure drop across the nozzle is of the same order as that across the tubes. With water only flowing through the nozzle, for a given water flow, the pressure drop is unaffected by variations in the heat input in the

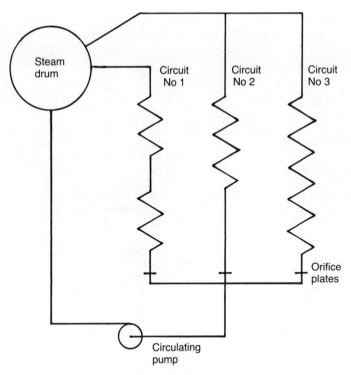

**5.10**   Diagram of a typical forced-circulation boiler circuit

tubes with which they are associated. The net result is that nozzles reduce
the effects that variations of heat input and tube configuration between
parallel circuits have upon the total system pressure drop and hence on the
flow through individual tubes.

  To ensure that the desired minimum water flow is achieved in all tubes,
when using natural circulation, it may be necessary for the flow in some
circuits within the boiler to be far in excess of the minimum required. The
improved stability of flow given to individual circuits by using forced
circulation enables somewhat lower total circulating water quantities to be
used than is the case with natural circulation. This, and the ability to use
higher pressure drops in the various parts of the circuit, enables less or
smaller diameter external circulating pipework to be used. The penalties of
forced circulation are increased boiler auxiliary power consumption and
additional equipment, e.g. pumps, on which maintenance will be required.

  Whether natural or forced circulation is used, a detailed analysis of the
heat input and water flow through all tubes is necessary to ensure that a
satisfactory water flow is achieved throughout the boiler. Where a number
of fuels are to be used, each mode of firing must be considered,
particularly where gases and liquids are used giving widely differing heat

absorption rates, see Table 15.5. Graphical and computerised procedures are of considerable assistance on all but the simplest of boilers.

# References

1. McAdams, W. H. *Heat Transmission* (3rd edn). McGraw-Hill.
2. Kern, D. Q. *Process Heat Transfer*. McGraw-Hill.
3. McAdams, W. H. *Heat Transmission* (3rd edn). McGraw-Hill, p. 82–98.
4. Ibid., p. 99–105.
5. Ibid., p. 72.
6. Gunn, D. C. *Heat Transfer in the Furnaces of Shell Type Boilers*. Proceedings Inst. Mech. Eng. Steam Plant Group. 187: 64/73.
7. McAdams, W. H. *Heat Transmission* (3rd edn). McGraw-Hill, p. 69.
8. Ibid., p. 88.
9. Ibid., p. 445–60.
10. Kern, D. Q. *Process Heat Transfer*. McGraw-Hill, p. 795–803.
11. Standards of Tubular Exchanger Manufacturers Association (TEMA), USA (4th edn), pp. 91, 92.
12. *Perrys Chemical Engineers Handbook*. McGraw-Hill, pp. 23–31 to 23–47.
13. McAdams, W. H. *Heat Transmission* (3rd edn). McGraw-Hill, p. 13.
14. Knudsen, J. G. and Katz, D. L. *fluid Dynamics and Heat Transfer*. McGraw-Hill, p. 3.
15. McAdams, W. H. *Heat Transmission* (3rd edn). McGraw-Hill, p. 219.
16. Ibid., p. 219.
17. Ibid., 272.
18. Grimison, E. D. *Trans. ASME* 59(1937) and 60(1938).
19. McAdams, W. H. *Heat Transmission* (3rd edn). McGraw-Hill, pp. 272–4.
20. Roderick, D. J. I., Murray, M. V and Wall, A. G. 'Heat Transfer and Draught Loss in the Tube Banks of Shell Boilers', *J. Inst. Fuel* 1959 (Oct.) **32**, 450.
21. BS 2790:1986 *Shell Boilers of Welded Construction*, British Standards Institution, London.
22. Kern, D. Q. *Process Heat Transfer*. McGraw-Hill, pp. 845–6.
23. TEMA, pp. 60–2.
24. McAdams, W. H. *Heat Transmission* (3rd edn). McGraw-Hill, p. 189.
25. Kern, D. Q. *Process Heat Transfer*. McGraw-Hill, p. 42.
26. Bowman, R. A., Mueller, A. C. and Nagle, W. *Trans. ASME*, **62** (1940).
27. McAdams, W. H. *Heat Transmission* (3rd edn), McGraw-Hill, pp. 194–5.
28. Kern D. Q. *Process Heat Transfer*. McGraw-Hill, p. 515.
29. McAdams, W. H. *Heat Transmission* (3rd edn). McGraw-Hill, p. 268.
30. Kern, D. Q. *Process Heat Transfer*. McGraw-Hill, p. 525 *et seq.*
31. *ESTUCO Welded Finned Tubes, Heat Transfer Design Data Book*. Extended Surface Tube Co.
32. ESCOA *Corporation Bulletin T-100a. Industrial Heat Transfer Equipment*. USA.
33. Evans, S. I. and Sarjant, R. J. 'Heat transfer and turbulence in gas flowing inside tubes'. *J. Inst. Fuel* 1951 (Sept.).
34. Spalding, D. B. 'Heat transfer aspects of coal science', Fourteenth coal science lecture, *BCURA Gazette* (1965) **55**.

35. BS 1113: *Design and Manufacture of Water-tube Steam Generating Plant (including superheaters, reheaters and steel tube economisers)*. British Standards Institution, London.
36. McAdams, W. H. *Heat Transmission* (3rd edn). McGraw-Hill, p. 145.
37. Kern, D. Q. *Process Heat Transfer*. McGraw-Hill, p. 52.
38. McAdams, W. H. *Heat Transmission* (3rd edn), McGraw-Hill, p. 158.
39. Moody, L. F. 'Friction factors for pipe flow', *Trans. ASME*, 1944 (Nov.) **66**: pp. 671–84.
40. McAdams, W.H. *Heat Transmission* (3rd edn). McGraw-Hill, p. 158.
41. Rose, J. W. and Cooper, J. R. *Technical Data on Fuel* (7th edn), 1977. British National Committee, World Energy Conference London, pp. 35–6.
42. McAdams, W. H. *Heat Transmission* (3rd edn). McGraw-Hill, p. 161.
43. Ibid., p. 162.
44. Ibid., pp. 162–4.

# CHAPTER 6

# Water treatment and feedwater systems

## 6.0 Introduction

The correct treatment of water for use in steam and hot-water boiler plant is an important contribution to extended trouble-free operation and should be fully appreciated by those associated with the design and operation of boilers. Modern highly rated boilers require more stringent water-side control than the older more moderately rated types which, as a result, were more tolerant of water conditions. This is an important consideration when a modern boiler is being installed alongside older designs, and revised water treatment may be necessary.

## 6.1 Sources of raw water

Raw water is a term used referring to water obtained from a source before receiving any treatment and is always unsuitable for use in boilers, either steam or hot-water. All natural water sources, be they rivers, springs, wells or the sea, contain impurities that have been acquired during the cycle to which they have been subjected. This cycle is shown on Fig. 6.1, which indicates that the atmospheric moisture is the result of evaporation from the various sources. When it condenses it falls as rain, hail or snow, absorbing gases from the atmosphere and also other material discharged by us into the atmosphere, including dust particles. Rainwater therefore contains a considerable quantity of impurities by the time it reaches the ground.

As the water flows over the surface of the earth or soaks through the earth's outer layer, further impurities are picked up or dissolved from the soil and minerals through which it passes. Spring and well waters, which are generally clear, still contain considerable dissolved solids.

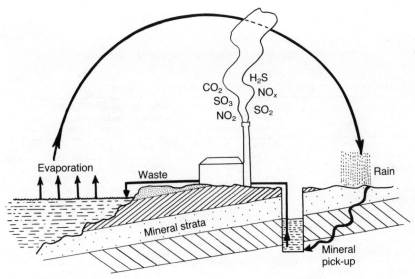

**6.1** Water cycle

Rivers contain impurities from the various effluents discharged into them on their way to the sea, including decaying vegetable matter, fertilisers and detergents, all of which can have undesirable effects on boilers. The impurities in well and spring waters tend to be of fairly constant composition, whereas those in rivers are subjected to seasonal variations. In countries with extreme variations of climate from drought to heavy rain, or where the river passes through a number of climatic areas, very variable dissolved and suspended solid contents can occur. As a consequence, to obtain meaningful analyses of the water from a river on which the design of a suitable treatment can be based, it is necessary to obtain frequent samples for at least one complete yearly cycle, preferably more, to ensure that details of extremes are known.

The extent to which water will dissolve minerals is also variable. Some are less soluble in pure water than they are in water which already contains other chemicals; for example silica, which can have serious effects in boilers, is more soluble in the presence of alkaline impurities than in pure water.

Impurities contained in raw waters can be some or all of the following.

1.  suspended solids and liquids not miscible with water, e.g. oil;
2.  colouring materials, particularly in moorland areas;
3.  bacteria and other micro-organisms;
4.  semi-colloidal substances;
5.  dissolved gases; and
6.  dissolved mineral salts,
    (*a*)  cations,

(*b*) anions, and

(*c*) silica.

Table 6.1 shows some typical examples of raw water analyses.

**6.1** Typical examples of types of raw water and their analyses

| | Moorland surface drainage | Lowland surface drainage | Deep well water | Industrial rain- water | Rural rain- water | Granite strata water | Chalk strata water |
|---|---|---|---|---|---|---|---|
| pH | 6.7 | 7.5 | 7.0 | 4.1 | 7.0 | 7.3 | 7.8 |
| Calcium (mg l$^{-1}$ Ca$^{++}$) | 7 | 63 | 48 | 7.2 | 13.6 | 5.7 | 100 |
| Magnesium (mg l$^{-1}$ Mg$^{++}$) | 6 | 18 | 20 | 1.5 | 1.5 | 0.8 | 3.6 |
| Sodium (mg l$^{-1}$ Na$^{+}$) | 8 | 25 | 74 | 4.6 | 6.9 | — | — |
| Potassium (mg l$^{-1}$ K$^{+}$) | — | — | — | — | — | — | — |
| Bicarbonate (mg l$^{-1}$ HCO$_3^{-}$) | 15 | 160 | 350 | Nil | 18 | 6 | 114 |
| Chloride (mg l$^{-1}$ Cl$^{-}$) | 10 | 43 | 48 | 10.6 | 10.6 | 7 | 40 |
| Sulphate (mg l$^{-1}$ SO$_4^{=}$) | 31 | 80 | 10 | 27 | 10.6 | 5 | 77 |
| Nitrate (mg l$^{-1}$ No$_3^{-}$) | Trace | 15 | Trace | — | — | 2 | 25 |
| Silica (mg l$^{-1}$ SiO$_2$) | 8 | 10 | 15 | — | — | 2.2 | 3 |
| Total hardness (mg l$^{-1}$ CaCo$_3$) | 42 | 232 | 203 | 24 | 40 | 17.6 | 270 |
| TDS (mg l$^{-1}$) | 77 | 333 | 387 | 40 | 60 | 36.6 | 400 |

### 6.1.1 Methods of expressing concentrations

Because the quantities of chemicals contained in water are usually relatively small, they are expressed as milligrams per litre (mg l$^{-1}$), parts per million (p.p.m.) or, for very small quantities, parts per billion (p.p.b.). When salts dissolve in water they dissociate into cations and anions. The metallic parts (calcium, sodium, magnesium, hydrogen, etc.), being cations carry positive electrical charges. The non-metallic parts (sulphur, chlorine, oxygen, etc.), carry negative charges. A water analysis can be expressed in two ways:

1. by giving the concentration of each individual cation and anion in mg l$^{-1}$ without showing the way in which they are combined with each other in the water; or

2. by expressing each constituent in mg l$^{-1}$ as a chemically equivalent amount of calcium carbonate. This procedure is used to simplify water treatment calculations.

The total solids content of treated water or feed and boiler water can be estimated by measuring the electrical conductivity of a sample. Pure water is a poor conductor of electricity, with the conductivity increasing as the solids content increases. It is measured in microsiemens per centimetre ($\mu$S cm$^{-1}$) or micromhos per centimetre, both at a specified temperature. (1 microsiemen cm$^{-1}$ is equal to 1 micromho cm$^{-1}$.) The conductivity of water depends upon factors such as the nature of the solids contained and the water temperature, therefore it can only be indicative of the water condition unless a particular range of conditions is calibrated by chemical analysis. An approximate indication of the total dissolved solids (TDS) in the water can be obtained by multiplying the conductivity in microsiemens $cm^{-1}$ or micromhos $cm^{-1}$, by 0.7.

## 6.2   Steam purity and carryover

### 6.2.1   Steam purity

The steam leaving a boiler is rarely perfectly pure. In general terms, all the dissolved salts in the boiler water vaporise to some extent, in superheated steam. The degree of volatility depends upon the type of salts present and the operating pressure of the boiler. Because of this, limits are imposed on the permitted levels of certain salts and, in particular silica, in the boiler water such that an acceptably low level of dissolved solids and silica is maintained in the steam. This is termed 'chemical carryover' as distinct from droplet entrainment, which is termed 'mechanical carryover', in which the small droplets contain the chemicals that are dissolved in the boiler water.

Steam purity is expressed as dryness or wetness in the case of boilers supplying saturated steam and in terms of total dissolved solids for boilers supplying superheated steam. The reduction of water in the steam with firetube boilers is achieved by the use of low steam offtake velocities from the water surface; a saturated steam dryness of about 98% is often assumed. The velocity of steam disengagement from the water surface should not exceed 60 mm $s^{-1}$ in firetube boilers. With water-tube boilers, where the drums are, for a given steam output, of smaller diameter, a separation system consisting of some combination of baffles, screens or cyclone separators is installed in the steam drum. With superheated boilers a steam purity of 1 mg $l^{-1}$ (1 p.p.m.) or better is aimed for, depending upon the requirements of the steam consumer.

External separators may be fitted into the steam pipework of firetube boilers and water-tube boilers where they are subjected to rapid steam offtake to give an improvement of the steam quality. The recommended quality of feed and boiler water for use at various operating pressures is given in the Standards[1,2,3], or is specified by the boilermaker, usually in the operating manuals. In very broad terms, the total dissolved solids in the boiler should be less than 3500 mg $l^{-1}$, but much will depend upon the composition of the dissolved matter. Tables 2 and 3 of Reference 2 are more explicit. The steam purity or quality can be impaired by 'priming' and 'foaming'.

### 6.2.2 Priming

Priming is the carryover of significant quantities of water with the steam. If large quantities or 'slugs' of water pass into the steam pipe they can have disastrous results on turbine blades since they will be travelling at the velocity of the steam, which may be as high as 50 m s$^{-1}$ (180 km h$^{-1}$). Priming can be caused by:

1. operating the boiler with a high water level;
2. operation at a pressure below design, which increases the volume and therefore the velocity of the steam (see Fig. 6.2); and
3. excessive steam demand, resulting in a reduction in boiler pressure and a rise of water level due to the increase in the volume of the steam below the water level.

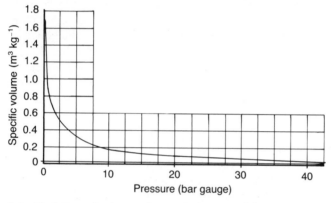

**6.2** Variation of volume of saturated steam with pressure

### 6.2.3 Foaming

Foaming caused by impurities in water is the formation of bubbles in the water at or above its surface which do not readily burst. As a consequence, they can build up and carry over with the steam, taking water and solids with them.[1] Figure 6.3 shows such an event taking place in a firetube boiler, the foam being drawn up into the steam offtake. Pure water does not foam. The steam bubbles formed in a boiler with the correct water conditions burst on reaching the water surface.

Foaming can be caused by the following boiler water conditions:

1. high suspended solids;
2. high alkalinity;
3. high total dissolved solids; and
4. contamination by detergents, oil and other organic matter.

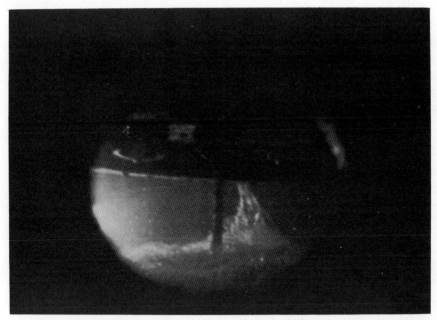

**6.3**   Foaming taking place in a horizontal packaged boiler. (Photographed through a quartz window in the rear tube plate. Foam is being drawn up towards the steam offtake which is behind the structure in the foreground. The stepped tube is a depth gauge and the boiler delivery is 1500 kW at 7.0 bar. This photograph is a 'still' from a research film, '*Boiling Phenomena in Boilers*', prepared by NEI Thompson Cochran Limited)

It can be overcome by the addition of antifoams, but the prevention of a foaming situation through correct water treatment is a more satisfactory solution than the use of antifoams which may not always be effective. Oil, in addition to causing foaming, can deposit on the water-side of boiler heating surfaces, insulating them from the cooling effect of the boiler water and hence causing overheating and failure. It should not therefore be allowed to enter the boiler.

## 6.3   Boiler blowdown

When water is boiled and steam is generated, any dissolved solids contained in the water remain in the boiler. If more solids are put in with the feedwater they will concentrate and may eventually reach a level where their solubility in the water is exceeded and they deposit from solution. If this occurs, the deposits can form in pockets throughout the boiler and eventually impair water circulation and boiler safety. To prevent such a situation arising, the boiler water solids content is limited (section 6.2.1), the level being controlled by the removal from the boiler of water with a high solids concentration.

This process of water removal from the boiler is called blowing down. It is carried out by the manual or automatic operation of a valve positioned so as to remove the water from the point of highest dissolved solids concentration. This water is discharged to a blowdown vessel or sump where flashing off and expansion to atmospheric pressure can be completed with safety, as illustrated in Fig. 6.6(*a*) (see also section 16.8.1).

The carryover of volatile substances in the boiler water, e.g. silica, is unaffected by the efficiency of the steam separation devices but, nevertheless, the amount carried over in the steam depends upon the concentration in the boiler water and hence it can also be controlled by blowdown.

Blowdown can be intermittent or continuous, or a combination of both, depending upon the quantity of water to be removed and replaced by feedwater. Wherever possible continuous blowdown should be used, there being a number of practical advantages.

1. The blowdown, and hence the quantity of feedwater necessary to replace it, is a steady flow instead of having short-term peaks.
2. The boiler water TDS is constant at a given load and hence so is steam purity.
3. Once the blowdown valve is set for given conditions there is no need for regular operator intervention.

Intermittent blowdown requires large short-term increases in the amount of feedwater put into the boiler and hence may necessitate larger feedwater pumps than if continuous blowdown is used. With intermittent blowdown the solids content of the boiler water is continuously varying, as illustrated by Fig. 6.4. This can create fluctuations of the water level in the boiler due to the changes in steam bubble size and distribution which accompany changes in concentration of solids.

In the case of watertube boilers, intermittent blowdown can affect the boiler circulation and hence boiler safety if taken from lower drums or headers.

Intermittent blowdown can be used to give a short-term high water flow which can move precipitates that have settled out in the lower parts of the boiler. This is of particular importance with firetube boilers where water treatment tends to be less rigorous than in water-tube boilers. If used for this purpose, intermittent blowdown should only be carried out at low load.

The quantity of blowdown required to control boiler water solids concentration is calculated by carrying out a mass balance of the solids to and from the boiler. The solids carried away in the steam leaving the boiler can be ignored at low pressures, where a high boiler water TDS is allowed,

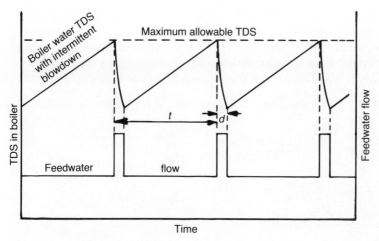

**6.4**    Effect of intermittent blowdown on boiler water TDS and feedwater flow

with an insignificant effect on the quantity of blowdown calculated. At high pressures, where the allowable TDS and silica in the boiler water are low, the solids in the steam can be significant to the calculation of blowdown quantity.

For continuous blowdown, with reference to Fig. 6.5:

$E$    is the boiler evaporation (kg h$^{-1}$);
$S$    is the solids concentration in the feedwater (mg l$^{-1}$); this is the

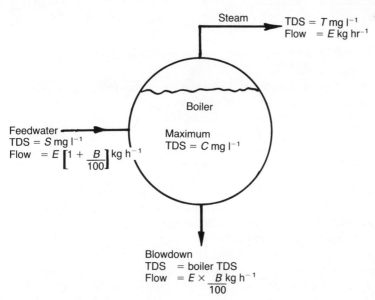

**6.5**    Total dissolved solids balance for a boiler

feedwater as fed to the boiler including solids in makeup, condensate return and any added chemicals;

$B$   is the continuous blowdown as a percentage of the boiler evaporation;
$C$   is the allowable solids concentration in the boiler (mg l$^{-1}$); and
$T$   is the solids concentration in the steam leaving the boiler (mg l$^{-1}$).

Solids entering with feedwater = solids leaving in blowdown + solids leaving in steam

$$E\left(1 + \frac{B}{100}\right)S = (E \times T) + \left(\frac{E \times B \times C}{100}\right)$$

hence

$$B = \frac{100\,(S - T)}{C - S}$$

If the solids content of the steam is neglected, i.e. $T = 0$

$$B = \frac{100S}{C - S} \tag{6.1}$$

For intermittent blowdown the calculation is more complex because the water content of the boiler and the amount of water blown down during each cycle have an influence on the results. By carrying out a mass balance of the solids entering and leaving the boiler with reference to Figs. 6.4 and 6.5 Equation [6.2] can be derived.

$$B = \frac{100\,SWt}{d\left(CW - \dfrac{ES}{2}\,(t - d) - SW\right)} \tag{6.2}$$

Where: $B$   is the blowdown rate as a percentage of the boiler evaporation;
$\quad\quad\quad S$   is the solids concentration in the feedwater (mg l$^{-1}$);
$\quad\quad\quad W$   is the water content of the boiler (kg);
$\quad\quad\quad E$   is the rate of boiler evaporation (kg h$^{-1}$);
$\quad\quad\quad C$   is the allowable solids concentration in the boiler (mg l$^{-1}$);
$\quad\quad\quad t$   is the interval between the commencement of consecutive blowdown operations (h); and
$\quad\quad\quad d$   is the duration of each blowdown operation (h).

This expression is best solved by trial and error using assumed values for two of the parameters $B$, $t$ and $d$ to calculate the third.
If $t = d$, i.e. blowdown is continuous, then

$$B = \left(\frac{100S}{C - S}\right)$$

as equation [6.1].

When calculating the blowdown, any chemicals injected into the feedwater and boiler which will result in an increase in the TDS of either must be taken into consideration. The quantity of blowdown can be dictated either by the TDS or by the silica content. Both should be checked using equation [6.1] or [6.2], substituting the silica contents for the TDS.

As the blowdown leaves the boiler it exists as saturated water at boiler pressure. At the discharge end of the blowdown pipe it consists of a mixture of steam and water, the steam having 'flashed off' as a result of the sensible heat released from the water when the pressure falls. The blowdown therefore has to be discharged into a blowdown vessel or sump such that the steam can be discharged to a point where it will cause no nuisance or danger to personnel. It is important that the steam discharge pipe from the vessel or sump is of ample size to prevent pressurisation of the sump or vessel.

It should be appreciated that because saturated water is blown down, useful heat that has been added as sensible heat to raise this water from feedwater temperature to saturation temperature is also discharged. This heat quantity is approximately 20% of that required to convert the same quantity of water to steam and hence can be significant with high blowdown rates.

If plant economy is of concern, heat recovery from the blowdown and, possibly, recovery of the flashed steam to conserve water may be justified. Heat recovery is achieved by passing the blowdown through a heat exchanger using boiler feedwater as the cooling medium as illustrated by Figs. 6.6(*a*) and 6.6(*b*). The latter arrangement necessitates the blowdown flash vessel operating at least at de-aerator operating pressure (see section 6.8). Recovery of heat and steam are usually only worth while with continuous blowdown.

## 6.4    Objectives of water treatment

Treatment of water for use in boiler plant is necessary to protect the boiler pressure parts in contact with the boiler water against corrosion and to prevent the formation of scale. This ensures that the plant is maintained in a satisfactory condition and also exerts some control over the TDS in the boiler water and hence reduces blowdown. To do this it is necessary to reduce, or preferably remove, the deposit- or scale-forming properties of the waters.

The formation of scale or deposit on part of a boiler heated by the products of combustion of the fuel will result in a rise of the metal temperature (Ch. 5) due to the insulating effect of the scale.[1] If this is allowed to progress it can cause deformation and ultimately failure of

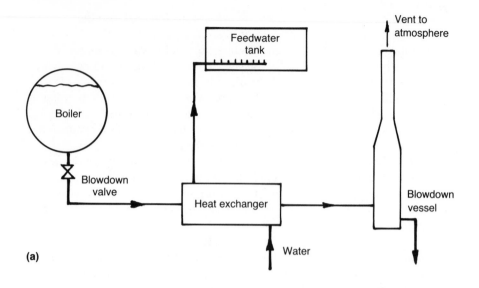

**(a)**

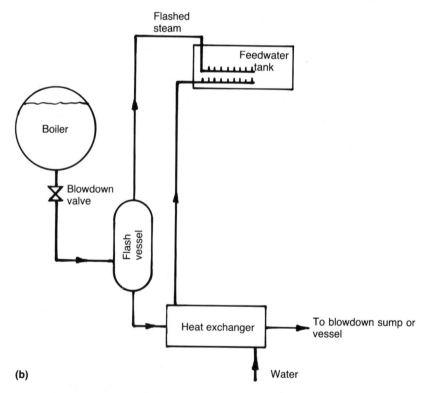

**(b)**

**6.6**  Blowdown system diagrams:
  (*a*)  with heat recovery; and
  (*b*)  with heat and flash steam recovery

heated parts due to the reduction of mechanical strength at high temperatures.

Without scale on the water side of the boiler the metal temperature is substantially the same as the boiler water temperature. In a steam boiler operating at 7 bar gauge the water temperature is 170 °C and at 30 bar gauge it is 236 °C.

With firetube boilers the furnace metal thickness can be up to 22 mm and the tube plates sometimes even thicker than this. Heat fluxes are high in these areas, in the range 150 to 300 kW m$^{-2}$, and locally even higher. There will be an appreciable temperature gradient through this metal which will set up thermal stresses. A layer of scale on the water side will tend to raise the metal temperature, in some cases into the creep range, where the metal, distorted by stress, will not recover to its original dimensions and the furnace may bulge. Subsequent heating and cooling will increase the distortion to such an extent that it may finally rupture, with disastrous results. It is important, therefore, that the formation of scale is minimised, preferably prevented entirely, which, with modern treatment methods carefully applied can be achieved. The effect of scale on metal temperatures at mid-thickness is illustrated by Fig. 6.7. This is for a scale of average thermal conductivity, e.g. a carbonate scale. Porous scales and silicate scales have very much lower conductivities and even a thickness below 1 mm can have severe effects. Table 6.2 presents the conductivities of some common boiler scales.

**6.2**  Conductivity of some common boiler scales

| Nature of scale | Thermal conductivity (W m$^{-1}$K$^{-1}$) |
| --- | --- |
| Calcium carbonate | 0.86 − 2.6 |
| Calcium sulphate | 0.69 − 1.3 |
| Silicate scale | 0.08 − 0.13 |
| Porous scales | 0.09 − 0.90 |
| Dense scales | 2.2  − 3.5 |

It is the physical rather than the chemical nature of the scale which affects thermal conductivity. Under working conditions a porous scale will entrap steam, the conductivity of which is very low.
From BCURA Information Circular No. 15 (British Coal)

Tube plates suffer differently from the effects of scale. They are tied together by stays which prevent gross bulging and eventual rupture, but the stresses can be such as to cause the tube seats to crack and leak. Scale on the water side of the ligaments between the tube seats is inaccessible for manual removal, chemical cleaning is the only way to remove it. Carbonate scales can be removed by inhibited acid but silicate scales are more difficult – a boil-out with hot caustic soda is one method (see section 6.13). Waters which are acid or very alkaline will attack the steel used in the construction of the boiler pressure parts, resulting in corrosion and eventual failure.

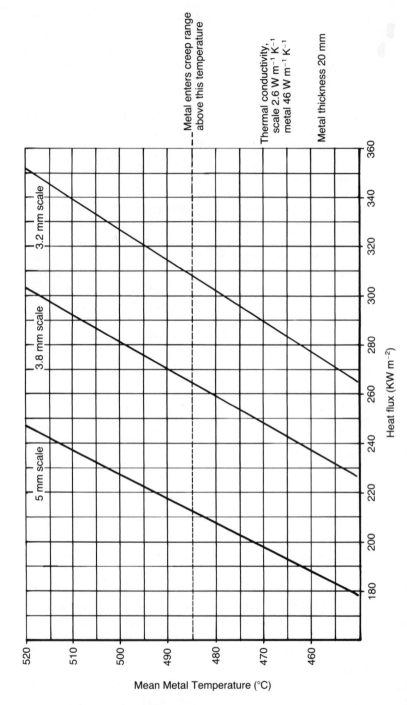

**6.7**   Effect of scale and heat flux on metal temperatures

## 6.5   Impurities causing deposits

### 6.5.1   Hardness

The most important chemicals contained in water that influence the formation of deposits in boilers are the salts of calcium and of magnesium, which are known as hardness salts.

Calcium and magnesium bicarbonate dissolve in water to form an alkaline solution, and hence these salts are known as alkaline hardness. Upon heating, they decompose, releasing carbon dioxide, the resulting chemicals settling out as a soft sludge. These used to be called temporary hardness, i.e. hardness that can be removed by boiling.

Calcium and magnesium sulphates, chlorides and nitrates, etc., when dissolved in water are chemically neutral and are known as non-alkaline hardness. These used to be called permanent hardness and form hard scales on boiler surfaces which are difficult to remove. Non-alkaline hardness chemicals fall out of solution due to reduction in solubility as the temperature rises, by concentration due to the evaporation which takes place within the boiler or by chemical change to a less soluble compound.

### 6.5.2   Silica

The presence of silica in boiler water can give rise to the formation of hard silicate scales, it can also associate with calcium and magnesium salts, forming calcium and magnesium silicates of very low thermal conductivity. Silica can give rise to deposits on steam turbine blading, the silica having been carried over either in droplets of boiler water in the steam or, at higher boiler pressures, by direct volatilisation of the silica into the steam.

## 6.6   Impurities causing corrosion

Corrosion is caused by electrolytic action set up by the dissolved gases oxygen, hydrogen or carbon dioxide in the water.

With oxygen and carbon dioxide the metal interacts with the gas and leads to the iron being dissolved or being converted to an insoluble oxide, the result being pitting and eventual failure of the pressure part. Often, internal corrosion takes place beneath porous scales where boiler water salts and chemicals concentrate and become corrosive, and metal wastage occurs. Figure 6.8 illustrates an example of this.

With hydrogen corrosion caused by acidic water conditions, the carbon in the steel reacts with the hydrogen, causing the collapse of the metal

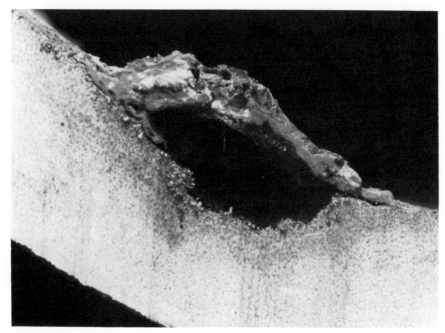

**6.8**   Example of a corrosion pit beneath the scale inside a tube of a water-tube boiler

structure so that it becomes brittle, i.e. hydrogen embrittlement. The characteristic of such failures is a thick-edged hole.

Caustic embrittlement is a form of corrosion occurring when steel, stressed beyond its yield point, comes into contact with strong sodium hydroxide or potassium hydroxide solution (greater than 5%, a strength much higher than should exist in boiler water). Caustic embrittlement is unlikely in modern welded boilers, but high concentrations of caustic hydroxides could occur at leaking tube expansions where the water flashes off, leaving a high concentration of caustic in the leakage path.

One form of 'acceptable' corrosion within boilers is the reaction between the iron and water which eventually results in the formation of a magnetite film ($Fe_3O_4$) which protects the boiler pressure parts against the electrolytic corrosion actions previously described. To maintain this protective film the water must be alkaline and have reducing conditions.

## 6.7   Water treatment processes

These can be either internal or external; both use chemicals to achieve their results.

## 6.7.1   Internal treatment

This is where chemicals are added to the boiler to prevent the formation of scale by converting the scale-forming compounds to free-flowing sludges which can be removed by blowdown.

Proprietary chemicals are available to suit the various water conditions. The specialists must be consulted to determine the most suitable chemicals to use in each case. This method of treatment is limited to boilers where the feedwater is low in hardness salts and, to low pressures where high total dissolved solids contents in the boiler water are tolerated. If these conditions are not applied, high rates of blowdown are required to dispose of the sludge. They then become uneconomical from heat and water loss considerations.

Internal treatment alone is not recommended. Boilers are not intended to be chemical reactors. Salts added to feedwater should be restricted to those needed to enhance the treatment after the water has been processed as far as possible before it enters the boiler, as subsequently discussed.

## 6.7.2   External treatment

This is where the water is treated by de-aeration and resin exchangers before it enters the boiler. Even with external treatment, some addition of chemicals to the feedwater is necessary to maintain the desired conditions and chemical residuals within the boiler.

The type of water treatment selected will depend upon the quality of raw water available and the quality required for the boiler. This, in turn, depends upon the boiler type and on the operating pressure (see section 6.2.1). A further consideration is the quantity of treated water or 'make up' required to replace losses from the boiler and steam system.

Boilers operating with some processes such as in sugar factories and with condensing turbines have a high proportion of clean, solid-free condensate returned for re-use in the boiler. In such cases, where make up may be less than 5% of the boiler output, a higher solids concentration in the make up may be tolerated to give an acceptable mixed feedwater TDS than would be acceptable in the case of 100% make up, i.e. no condensate returns.

The overall capital and operating costs of water treatment need to be investigated to determine the optimum solution for a particular application; this is a job for water treatment specialists.

The external treatment processes available are: lime softeners; lime/soda softeners; base exchange de-alkalisation; demineralisation; distillation; reverse osmosis; and de-aeration. These produce varying qualities of treated water and can be used individually or in various combinations to suit the composition of a particular raw water. Before any of these are used it may be necessary to remove suspended solids and colour from the raw water, particularly if taken from rivers, because these can foul the resins used in the subsequent treatment plant.

Methods of pre-treatment include simple sedimentation in settling tanks or the use of clarifiers with coagulants and flocculants being added if necessary to assist settling. Pressure sand filters with spray aeration to remove carbon dioxide and iron may be used to remove metal salts from deep well waters.

The first stage of treatment is to remove hardness salts and, possibly, non-hardness salts. Removal only of hardness salts is called softening, while total removal of salts from solution is known as demineralisation.

Lime softening precipitates alkaline (temporary) hardness salts by converting the calcium bicarbonate to calcium carbonate and the magnesium bicarbonate to magnesium hydroxide.

Lime/soda softening reduces both alkaline and non-alkaline (permanent) hardness. This may be sufficient treatment for low-pressure boiler plant. The lime/soda process is also employed as pre-treatment for an ion-exchange plant to reduce the loading on the resins.

The ion-exchange processes use special resins in the form of small beads which are capable of exchanging ions with those of the salts in the water. The simplest form is known as 'base exchange' or 'sodium exchange' (Fig. 6.9), in which the calcium and magnesium ions are exchanged for sodium. The sodium salts, being soluble, tend not to form scales in boilers. When all the sodium in the resin is exchanged, the resins become exhausted and 'break-through' occurs, i.e. raw water passes through unchanged. It then becomes necessary to regenerate the resin by replacing the sodium. This is done by washing with strong brine solution (sodium chloride).

Because base exchange only replaces the calcium and magnesium with sodium, it does not reduce the total dissolved solids content, and high

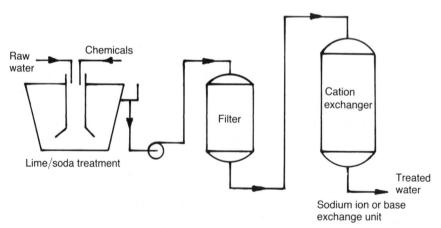

**6.9**   Base exchange preceded by lime/soda softener,
(From BS 2486 (1978) Fig. 8)

blowdown rates can be necessary with consequent high make-up quantities. Base exchange does not reduce the alkalinity.

If a high make up is required it may be necessary to remove this alkalinity, and this can be achieved by a de-alkalisation process placed before the base exchange unit.

De-alkalisation is an ion-exchange process in which the hydrogen ions from the resin replace the calcium and magnesium associated with the bicarbonate, leaving a solution of carbon dioxide in water. The carbon dioxide is removed by passing the water through a degassing tower, dosing of the degassed water with caustic soda being necessary to raise the pH. Regeneration is by dilute sulphuric or hydrochloric acids. When followed by base exchange, this converts the non-alkaline hardness to sodium salts. The resulting treated water is lower in total dissolved solids than is the raw water, due to the removal of the alkaline hardness salts.

Demineralisation is the complete removal of all salts; the process is illustrated diagrammatically in Fig. 6.10. This is achieved by using a 'cation' resin which exchanges the cations in the raw water with hydrogen ions, producing hydrochloric, sulphuric and carbonic acids, the latter being removed in a degassing tower in which air is blown through the acid water. Following this, the water passes through an 'anion' resin which exchanges anions with the mineral acids and forms water according to the following reaction.

anion resin + sulphuric acid $\longrightarrow$ resin after exchange + water

$R_4NO\overline{H} + H_2SO_4 \longrightarrow R_4NHSO_4 + H_2O$

When the two resins are mixed in the same vessel this is known as a 'mixed bed'.

Regeneration of the cations and anions is necessary at intervals using,

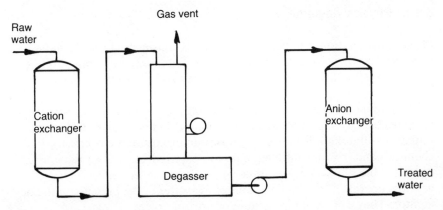

**6.10**   Demineralisation process
From NEI Thompson (Kennicott) Limited)

typically, mineral acids and caustic soda respectively. The complete removal of silica can be achieved by the correct choice of anion resin.

Reverse osmosis uses the fact that when solutions of differing concentrations are separated by a semi-permeable membrane, water from the less concentrated solution passes through the membrane to dilute the liquid of high concentration. If the solution of high concentration is pressurised, the process is reversed and the water from the solution of high concentration flows to the weaker solution. This is known as reverse osmosis. The quality of the water produced depends upon the concentration of the solution on the high-pressure side and the pressure differential across the membrane. This process is suitable for waters with very high TDS, such as sea water.

Distillation or evaporation is now largely replaced by ion exchange and reverse osmosis, which have lower operating costs. Distillation requires a supply of heat to evaporate the water and a cooling medium to condense it, hence the higher operating costs.

## 6.8 De-aeration

As stated in section 6.6 the main cause of internal corrosion of boiler pressure parts is the gases dissolved in the feedwater. To prevent corrosion and to preserve the protective magnetite film on the inner surfaces of the boiler, it is necessary to have reducing and alkaline conditions. To ensure that reducing conditions exist, all traces of oxygen must be removed from the boiler water. This can be done by external physical de-aeration, by the introduction of chemicals, or by both together.

### 6.8.1 Physical de-aeration

Physical de-aeration is to be preferred and is considered by some to be necessary at all pressures above 28 bar gauge and essential above 70 bar gauge and on occasions where there is a very high percentage make up of aerated treated water.

The initial cost of a de-aerator is high but there is little in the way of running costs. De-aeration by the addition of chemicals gives minimum initial costs but there is the continuous running cost of the necessary chemicals and blowdown to maintain the required TDS. Over the normal operating life physical de-aeration can be the most economical. De-aerators operate at the boiling point of the water at the pressure in the de-aerator. They can be of the vacuum or pressure type.

The vacuum type of de-aerator operates below atmospheric pressure, at

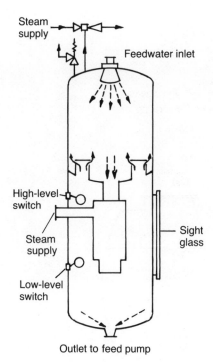

Steam supply

Feedwater inlet

High-level switch

Steam supply

Low-level switch

Sight glass

Outlet to feed pump

**6.11**   Vacuum-type de-aerator
(From NEI Thompson (Kennicott) Limited)

about 82 °C, can reduce the oxygen content in the water to less than
0.02 mg l$^{-1}$, but they require vacuum pumps or steam ejectors to maintain
the vacuum (Fig. 6.11).

Pressure-type de-aerators can reduce the oxygen content to 0.005 mg l$^{-1}$
and require only a pressure control valve on the steam supply to maintain
the desired operating pressure and hence temperature at a minimum of
105 °C (Fig. 6.12).

Where excess low-pressure steam is available, the operating pressure can
be selected to make use of this steam and hence improve heat economy.
Steam-driven feedwater pumps can be arranged to exhaust at a pressure
sufficient to supply the de-aerator, although a situation can occur at low
loads where the steam supply may exceed the demand.

## 6.8.2   Chemical de-aeration

De-aeration by the introduction of chemicals is achieved by the injection of
'oxygen scavengers', which may be catalysed sodium sulphite or hydrazine,
into the feedwater on the suction side of the feedwater pumps and which is

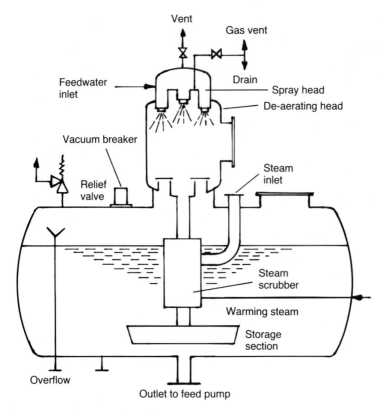

**6.12**  Pressure-type de-aerator
(From NEI Thompson (Kennicott) Limited)

referred to as low-pressure dosing. Even when using physical de-aeration, one of these chemicals is injected, albeit in much smaller quantities, to dispose of any residual oxygen after de-aeration. A small excess of chemical is maintained in the boiler as a safety measure.[2]

Sodium sulphite reacts with oxygen to form sodium sulphate which increases the TDS in the boiler water and hence increases the blowdown requirements and make-up water quantity.

Hydrazine reacts with the oxygen to form nitrogen and water. It is invariably used at high pressures when low boiler water solids are necessary, as it does not increase the TDS of the boiler water. It is also a reducing agent and reacts with iron oxide ($Fe_2O_3$) to form the protective magnetite film ($Fe_3O_4$). Hydrazine is highly toxic and care must be exercised when it is being handled. It should not be used if there is any risk of the steam from the boiler coming into contact with food or drink as may be possible in certain process industries.

## 6.9    Chemical dosing

### 6.9.1    Low-pressure dosing

As already stated (in section 6.8.2), low-pressure dosing of sodium sulphite or hydrazine is used with or without de-aeration for oxygen control.

Sodium hydroxide (caustic soda) is also introduced into the system at low pressure, at which it is necessary to maintain alkaline conditions, i.e. the required pH, in the boiler to prevent hydrogen corrosion occurring and to maintain silica in solution. Excess caustic must be avoided, otherwise caustic embrittlement may occur (see section 6.6).

### 6.9.2    High-pressure dosing

High-pressure (HP) dosing of trisodium phosphate is introduced direct to the boiler to provide an excess of phosphate ions that will combine with any residual hardness salts introduced into the boiler. Where a high alkalinity exists other forms of phosphates can be introduced to effect a reduction of this. Phosphate can be added into the feed pump suction, although this is not recommended when economisers are fitted. Phosphates can precipitate out in the economiser tubes when the water temperature rises, eventually causing blockage. This can be a problem if economiser recirculation is incorporated using water from the boiler to heat the incoming feedwater to the desired temperature (see section 7.4.1.2).

Where bicarbonates and carbonates have not been completely removed by water treatment, there will be carbon dioxide released in the steam, which will cause corrosion of the condensate system. If oxygen is also present, the corrosion can be considerably worse. To combat this corrosion, volatile amines are added to the feedwater. They distil over with the steam from the boiler and neutralise the carbon dioxide. Compounds in common use are morphaline and cyclohexylamine. The amounts required depend upon the amount of $CO_2$ generated. They should be continually injected and may be mixed with, and added along with, other low-pressure dosing chemicals.

Where the carbon dioxide concentrations are high, filming amines are available, which are injected into the steam system and form a protective film on the condensate system, thus preventing corrosion. The quantity used is independent of the amount of corrosive constituent present.

When a boiler is fitted with a controlled superheater (section 2.5.1), care has to be taken with the selection of the type of desuperheater to use. With direct-injection spray types, the use of boiler feedwater as the cooling medium can introduce solids into the steam, hence contaminating the relatively pure steam being supplied from the boiler. If spray types are used, the feedwater must have a total solids content of less than 1 mg l$^{-1}$, using hydrazine for low-pressure dosing if necessary. If this is not practicable due to a restriction on the use of hydrazine, a form of desuperheater which avoids direct contact between steam and cooling water should be used.

## 6.10 Establishing the capacity of feedwater treatment plant and feed systems

### 6.10.1 Steam and water losses

When determining the quantity of make-up water required and hence the capacity of the water treatment plant to install, all losses of steam and water from the complete water and steam system must be established. Losses include:

1. steam used for soot blowing;
2. water lost to blowdown;
3. steam used for oil heating if the condensate is not recovered because of the risk of contamination;
4. steam not returning as condensate from the steam consumers;
5. water used for regeneration and backwashing of water-treatment plant; and
6. a nominal quantity for incidental losses, say 2½%.

Where a single stream of water treatment is being considered, sufficient water will have to be produced and stored during the operating period to supply the requirements for make up during the regeneration period and for any other anticipated water-treatment plant shut down for maintenance, etc. If multiple streams including standby are used, this latter point may be of no consequence.

Also to be considered is the capacity of the treated water storage. This has to hold sufficient water to supply the requirements for backwashing and to supply boiler feed make up during the regeneration of a single stream.

Soot blowing and intermittent blowdown can use relatively larger rates of water flow for a short period of time and during this period may well exceed the average treated water flow rate. To ensure that sufficient treated water is available to cover such occasions it is worth while producing a diagram showing how the net quantity of treated water in storage varies with time over a full water-treatment plant cycle from the end of one regeneration to the end of the next.

Consideration should also be given to the production and storage of · sufficient water to supply the boiler requirements during commissioning and after a maintenance or repair shut down when the boiler has been drained. Where small make-up quantities are required, it could take several hours to treat sufficient water to fill a boiler. In the case of hot water installations where the make up is small (but not negligible), filling the boiler and heating system may take many days since treated water must be used. In such cases, for example district heating schemes, consideration should be given to charging the boiler and system with demineralised water obtained by clean tanker from a nearby power station.

### 6.10.2 Feedwater pump capacity

Care must also be taken to ensure that any pumps associated with the feedwater system have sufficient capacity to cover blowdown quantities, etc., particularly if such flows are intermittent.

An example affecting the capacity of the feedwater pumps is the case where a de-aerator is incorporated in the system using steam direct from the boiler or boilers for heating. This steam will have to be supplied in addition to the amount the consumer requires for process use, and hence the capacity of the feedwater pumps will have to be increased to include this. This can be illustrated by an example where a net actual steam flow to process of 10 t $hr^{-1}$ is required from a boiler at 20 bar gauge saturated when supplied with feedwater at 15 °C. A de-aerator is included in the feedwater system to raise the feedwater to 105 °C using steam from the boiler at 20 bar gauge saturated for heating. Neglecting heat losses from the de-aerator and associated pipework, a heat balance for the de-aerator shows that 1.6 t $hr^{-1}$ of steam are required by the de-aerator to raise the feedwater to 105 °C from 15 °C. The gross boiler output has to be 11.6 t $hr^{-1}$ and the feedwater pump has to be of sufficient capacity to supply this plus any blowdown, etc.

It must be appreciated that, neglecting de-aerator heat losses to atmosphere, such a system does not change the heat put into the steam by the boiler. The 11.6 t $hr^{-1}$ is being raised in the boiler from feedwater at 105 °C and requires the same quantity of heat as 10 t $hr^{-1}$ from feedwater at 15 °C.

## 6.11   Feed and boiler water testing

Once the feed and boiler water control parameters are established for a contract, these will usually be included in the boiler operating manuals supplied to the user by the boiler manufacturer. It is then the responsibility of the user to maintain the water conditions within the limits set down to ensure maximum availability of and life from the boiler. This necessitates regular analytical testing of the feed and boiler water conditions. The boilermaker's recommendations should always be carried out. In the absence of adequate laboratory conditions on site, water treatment contractors will often be able to supply test kits and to train boiler operatives in their use. Tables 2 and 3 of Reference 2 give guidance on water conditions needed for firetube and water-tube boilers respectively. This Standard should be read in its entirety as a supplement to this chapter.

## 6.12    Off-load storage of boilers – water-side protection

If a boiler is taken off load for any length of time and internal inspection and cleaning is not actually being carried out, it is essential that the pressure parts are protected against corrosion, by the correct storage procedure. Storage can be wet or dry.

### 6.12.1    Wet storage

With hot-water boilers, the complete hot-water system including the boiler should be maintained completely full of correctly treated water.

For short- to medium-length shutdowns of up to, say, three months' duration, wet storage is used for steam generators. In these cases, the boiler, including the superheater when one is fitted, is completely filled with de-aerated water. This should contain sulphite or hydrazine in sufficient quantity to react with any oxygen in the water and have a pH of 10–11 maintained by the addition of ammonia. These conditions should be checked periodically and corrected as necessary.

The use of hydrazine and ammonia are preferred since, being volatile and not increasing the solids content, there will be no settling out of salts during storage which would necessitate time-consuming flushing before putting the boiler back on load.

Care must be taken to ensure that the water and chemicals within the boiler are thoroughly mixed by a circulating pump if necessary. Provision must also be made to allow for expansion and contraction due to temperature changes.

Protection against freezing should be considered, particularly with outdoor installations. This can be by the introduction of a heating coil at a suitable position in the boiler.

### 6.12.2    Dry storage

For extended outages the boiler should be completely drained, cleaned and dried out if necessary using a hot-air drier. The boiler can be maintained dry by placing bags or trays of a dessicant such as silica gel in each drum of water-tube boilers, or the boiler shell of firetube boilers. It should then be sealed and maintained airtight. Regular inspection of the dessicant should be carried out, replacing it as necessary. As an alternative, hot air can be blown continuously through the boiler or an electrical heater can be inserted.

Particular attention must be paid to any naturally undrainable areas such as pendant superheaters (see section 13.5), and any pipework associated with instruments and mountings. Compressed air can be blown through the tubes to remove the water, followed by hot air for drying.

## 6.13    Boiler water-side cleaning

### 6.13.1    Chemical cleaning

This refers to the cleaning of the water side of a boiler which is carried out, as part of the commissioning process, after a major repair/overhaul or if there has been an internal buildup of scale or other compounds in the boiler due to a relapse in the water-treatment process.

Chemical cleaning is carried out when commissioning a new boiler to remove oil, grease and other debris left behind during construction, and in particular during tube expansion on water-tube boilers. This is usually an alkaline boil-out using, typically, combinations of caustic soda, calcium carbonate and trisodium phosphate or other proprietary chemicals, which are added in proportion to the water content of the boiler, the various quantities being stipulated by the boilermaker. The boiler has to be fired to raise the temperature of the metal and solution to a level where the oil and grease will be removed and the solution will circulate around the boiler. After the boil-out, all sediment and material that has settled out in drums and headers must be removed and the boiler flushed out with clean water.

Acid cleaning carried out during commissioning to remove mill scale and silica compounds from the pressure parts is normally only used on boilers designed to operate at pressures above 64 bar, where boiler water requirements are more stringent. Typically, a dilute acid solution is used with a suitable inhibitor to protect the boiler surfaces against acid corrosion. The solution has to be pumped around the circuits to thoroughly clean the whole boiler. This is not easy on a natural-circulation water-tube boiler where there are multiple parallel water circuits. This work should be carried out by competent specialists who will have the necessary equipment and effluent disposal facilities available.

Removal of scales, etc., formed during operation, can be performed mechanically with rotary cleaners, and using flexible drives if the boiler design allows this. Modern high-pressure boilers using welded tubes have limited access, particularly to combustion chamber and economiser tubes, hence the only solution is chemical cleaning. The chemicals used depend upon the composition of the deposits. Competent specialists will determine the chemicals to be used and carry out the required procedure. Chemical cleaning is expensive and inconvenient. The only satisfactory answer is therefore to prevent the occurrence of scale by adopting proper water treatment.

Deposits in superheaters which result from carryover or priming can normally be removed by flushing with water.

### 6.13.2 Steam blowing

Where steam is passing to a turbine it is usual to clean the superheater and steam pipework during commissioning by steam blowing. To carry this out, the boiler is fired and steam blown through the superheater and pipework up to the turbine at a velocity above the normal MCR value by discharging at a low pressure. The steam discharges to atmosphere against a polished steel target plate and the process is repeated using clean plates until no scoring of the plate occurs, signifying that all loose scale and weld metal has been removed.

It may be necessary to provide special pipework to ensure that the steam is discharged safely and without nuisance.

Complex piping arrangements incorporating multiple boilers and steam consumers require careful consideration to ensure that all routes are correctly cleaned.

## References

1. *The Treatment of Water for Shell Boilers*. Association of Shell Boilermakers.
2. BS 2486: *Recommendations for Treatment of Water for Land Boilers*, British Standards Institution, London.
3. V. G. B. Kraftwerkstechnik–Mitteilungen der VGB. (Apr.)1972.

# CHAPTER 7

# Auxiliary equipment

## 7.0  Introduction

Boilers need auxiliary plant to enable them to operate. This includes:

1.  pumps, to supply water to the boiler from the feed supply for steam boilers or from the water return of a hot-water heating system;
2.  fans, to supply air to the combustion system, and to remove products of combustion from the boiler and deliver these to the chimney;
3.  chimneys, to discharge the products of combustion to the atmosphere under acceptable environmental conditions; and
4.  heat exchangers, to recover heat from the products of combustion and return it to the boiler system.

It is the purpose of this chapter to describe the principles involved in the auxiliaries and to draw attention to some of the problems that can arise.

A number of these items are exposed to the products of combustion of the fuel fired in the boiler and may therefore be subject to low-temperature corrosion (see Ch. 10). Methods of combatting this are referred to in this chapter.

## 7.1  Pumps

### 7.1.1  Steam boilers

For steam boilers the pumps need to raise the pressure of the feedwater to a value somewhat greater than the maximum operating pressure of the boiler and be capable of delivering a quantity in excess of the maximum evaporation of which the boiler is capable. This margin is 15–25% of the maximum continuous rating (MCR) quantity which, it is necessary because:

1. boilers need to be blown down from time to time, or continuously, to remove dissolved and suspended solids which concentrate as the water evaporates (Ch. 6);
2. under unfavourable, and usually avoidable, conditions priming, or carry over of water with the steam can occur, thus removing water from the boiler;
3. under some circumstances, excessive steam demand can occur; and
4. pump performance may deteriorate with time, particularly in remote areas where maintenance and feedwater quality are not carefully controlled. In such circumstances higher margins should be used. This situation is unusual with modern pump designs and well-maintained plant.

An important aspect of feed pump installation is to ensure that there is a net positive suction head (NPSH) available at the pump inlet, and to ensure that at no point in the suction pipework is it possible to create a pressure below the saturation pressure of the feedwater. This is particularly critical at the inlet to high-speed pumps and with hot feed (as it should be), otherwise vaporisation of the water can occur with consequent cavitation in the pump, indicated by a high-pitched squeal. This causes flow to be greatly reduced and, if allowed to persist, damage to the pump. To ensure adequate NPSH, the pipework leading to the pump inlet should be generously sized, giving water velocities of about $1 \text{ m s}^{-1}$, as straight as possible, and the feed tank or de-aerator should be elevated to a high position in the boiler house to give the desired pressure at the pump inlet branch. The pump manufacturer will indicate the minimum NPSH required for pump safety and, if in doubt, he should be consulted about the proposed suction pipework arrangement.

Various types of feed pumps, positive displacement, single-stage regenerative, single-stage centrifugal and multi-stage centrifugal have been used in the past, but modern practice depends on the multi-stage centrifugal type which provides the pressure/volume characteristics most favourable to boiler operation. Figure 7.1 illustrates a typical characteristic. It will be noted that as flow increases, the pressure generated by the pump decreases. In choosing a pump, therefore, it is important to ensure that the pressure provided at the maximum flow conditions, e.g. MCR + 20%, is above the maximum working pressure of the boiler, with a margin to cover pressure loss through check valves, control valves, economisers when fitted, and the feedwater delivery system to the boiler. Figure 7.1 also illustrates how the water pressure required at the boiler increases with flow due to the increasing pressure drop through the feedwater system and superheater, if fitted.

With firetube installations the feed pumps are usually electrically driven and mounted on the same base as the boiler, they are therefore peculiar to that boiler. On/off water level control may be used on small boilers, but

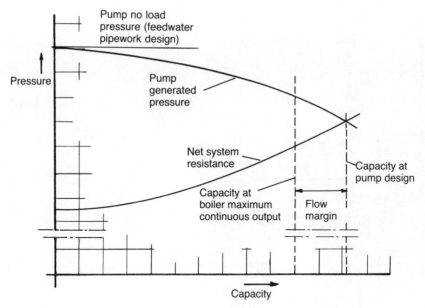

**7.1**  Typical characteristic curve for a centrifugal pump

modulating control is preferable and should be used where steam demand is subject to fluctuation, and with larger boilers.

A section through a pump of the type used extensively on firetube boilers is illustrated by Fig. 7.2. In special applications feed pumps are duplicated to ensure absolute reliability of water supply in the unlikely event of a pump failure. The impellors are made from stainless steel and thus have a long life.

With water-tube boilers, where more than one boiler is usually installed, the feedwater pumps are grouped together and always include standby capacity. They can be steam or electrically driven or a mixture of both and the operating pumps are usually running continuously, the boiler water level being controlled by a modulating valve (section 8.5.3.). The pressures to which water-tube boilers are applied necessitate multi-stage pumps which, because of their size and the power required to drive them, are usually of the horizontal-spindle type. (see Fig. 7.3).

Should a continuously running pump operate against a closed discharge valve with no flow through it, the water temperature will rise because of the energy input to the water by the driver. When using a modulating control with continuously running pumps, it is therefore necessary to ensure that there is always a minimum water flow through the pump to carry away the heat generated by the energy input. This is achieved by means of 'leak off', which can be a continuous flow from the pump discharge or by an automatic leak-off valve that operates only at low flow. A typical automatic leak-off valve is a combined non-return and leak-off

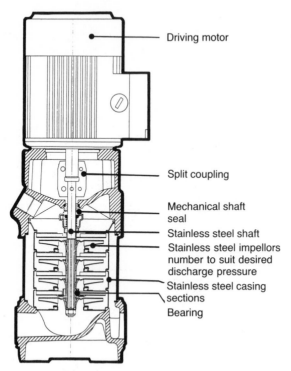

Driving motor

Split coupling

Mechanical shaft
seal

Stainless steel shaft

Stainless steel impellors
number to suit desired
discharge pressure

Stainless steel casing
sections

Bearing

**7.2**   Vertical-spindle multi-stage centrifugal feedwater pump
(From Grundfos Pumps Ltd)

valve where the leak off opens as the non-return valve closes on reduction
of flow to the boilers. The leak off is returned to the de-aerator or
feedwater tank.

### 7.1.2   Hot-water boilers

Hot water rather than steam is now quite extensively used for process work
as well as for space heating. The installation size can be quite large, for
instance in district heating plant and in certain breweries. The pumps
circulate water from the system return, through the boiler, through the
system and back to the pump inlet. The duty is therefore to circulate the
required volume of water, raising its pressure only sufficiently to overcome
the resistance of the system and the boiler rather than from atmospheric to
rather more than boiler pressure as is the case with boiler feed pumps.

   Centrifugal pumps are used for this purpose, but with fewer stages than
in boiler feed pumps, a single stage often sufficing. To allow for varying
demand, and hence flow rates, several pumps are used in parallel, but
more recent practice is to install one pump on each boiler. Subsidiary

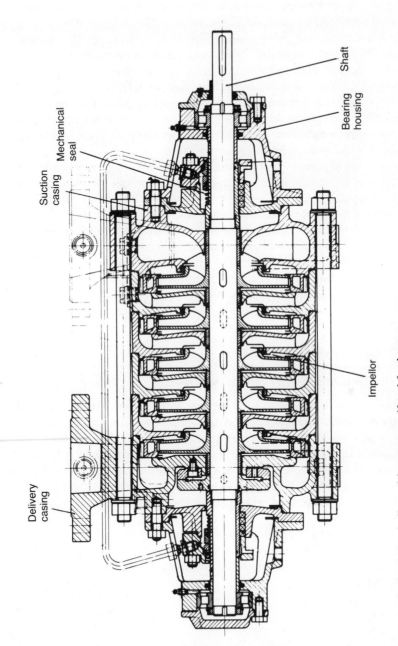

**7.3** Horizontal-spindle multi-stage centrifugal feedwater pump
(From SIHI–Ryaland Pumps Limited)

pumps may be necessary to supply different parts of a complex system, but these are not part of the boiler.

Most large hot-water systems operate at high pressure, the pressure being maintained by external pressurisation. This is applied by continuously running pumps which deliver water from storage vessels at atmospheric pressure to the boiler, which may be operating at 7 bar or more. Alternatively, the pressure can be applied from a small pressure vessel which contains water from the system and, in the space above the water surface, nitrogen introduced from ordinary gas cylinders at a controlled pressure. As the water expands or contracts with changes in temperature, water from the system is either discharged into, or pumped from, storage tanks at atmospheric pressure.

For both continuous pumping and nitrogen pressurisation, multi-stage centrifugal pumps are used, which are similar to boiler feed pumps but of smaller capacity. A centrifugal pump running at a given speed will deliver the same volume of any liquid to the same height irrespective of the density. Hence, it is usual to produce the characteristics for centrifugal pumps on a volume/head basis. If the head is converted to pressure, this varies with density, and hence cold water will be delivered at a higher pressure than hot water. The power required to drive the pump will be higher with higher density fluids.

There are three rules that can be useful in assessing the performance of a pump at varying speeds.

| | |
|---|---|
| Flow is proportional to speed | $V \propto N$ |
| Head is proportional to (speed)$^2$ | $H \propto N^2$ |
| Power absorbed is proportional to (speed)$^3$ | $kW \propto N^3$ |

The system resistance to be overcome by the pump can be determined by means of the vast amount of data available on this subject.[1] Care must be taken to cover the pressure drop of all valves and fittings, flow measuring devices and feedwater heaters. The difference in static head between the water level of the tank from which the pump takes the water, to the boiler water level must also be included. In some cases this may be a negative value if the feed tank is above the boiler, as it may be on firetube installations.

When designing the high-pressure feedwater pipework, the pump no-load (no-flow) pressure must be used (see Fig. 7.1.). This will be about 10% above the pressure generated at the maximum output (flow) condition.

## 7.2   Fans

These are devices used to move air and gases at relatively low pressures. They consist of a bladed rotor, wheel or impellor rotating in a housing, casing or volute which transfers energy to the gas or air in the form of velocity and pressure. There are two major types of fan:

1. centrifugal fans in which the fluid flow is basically radial and the centrifugal force acting on the air generates the pressure; and
2. axial flow fans which are of the propeller type enclosed in a housing with air flowing axially.

The fans most frequently used for industrial boiler work are of the centrifugal type.

A fan through which air is flowing converts the power put into the fan to kinetic energy and pressure energy. The actual, or static, pressure generated is measured by a manometer or other sensitive pressure gauge.

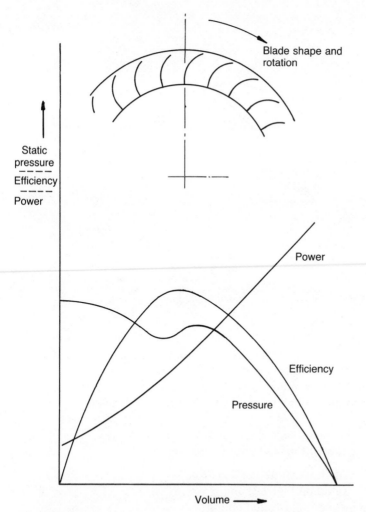

**7.4**   Typical characteristic curve of a forward-bladed fan

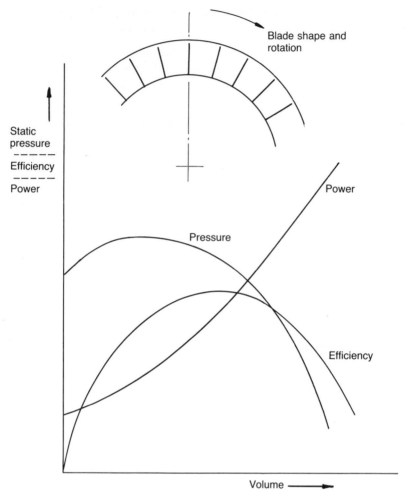

**7.5** Typical characteristic curve of a radial-bladed fan

The kinetic energy is measured by using a pitot tube pointing into the gas stream. The total pressure generated is the sum of the static plus velocity pressures. The pressure developed by a fan is relatively low and is measured in millibar or millimetres of water gauge (mm wg), the range of normal applications varying from zero to the order of 75 mbar (760 mm wg).

There are three basic blade shapes in use for centrifugal fan impellors, the shape being that at the blade tip. These are:

1. forward inclined, illustrated on Fig. 7.4;
2. radial, illustrated on Fig. 7.5; and
3. backward inclined, illustrated on Fig. 7.6.

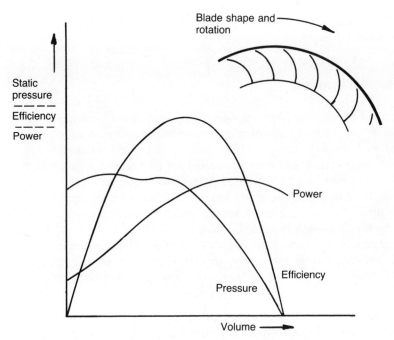

**7.6**    Typical characteristic curve of a backward-bladed fan

There are intermediate variations of these, with manufacturers introducing their own blade forms.

The relationship between the volume of gas or air discharged by a fan and the static pressure at which it is discharged is indicated by the above-mentioned illustrations. They all generate a high pressure at zero flow, i.e. with the discharge closed off. The pressure first rises to a peak as the flow is increased and then ultimately reduces to zero at maximum flow, i.e. with the discharge wide open.

### 7.2.1.  Forward-bladed fans

Forward-bladed fans deliver relatively high volumes and pressures for a given speed when compared with other blade shapes. They are of low to medium efficiency, producing the characteristic curves illustrated in Fig. 7.4. With increasing volume the pressure first falls slightly then rises again before falling to zero at maximum flow. The maximum efficiency occurs at or below the volume at which maximum pressure is generated. Forward-bladed fans may have multiple blades of little strength and hence should not be used at high speeds or with dust-laden gases.

Deposits from dirty gases tend to accumulate readily on the blades. This type of blade is ideal for handling large volumes of clean air or gas but it does tend to be noisy. Such blades are mostly used on the smaller firetube boiler for forced draught production.

### 7.2.2 Radial-bladed fans

Radial-bladed fans have a blade shape which permits robust construction, hence giving the capability to withstand erosion for a relatively long period compared with other blade types. They are therefore ideal for applications with a high particulate loading such as exhauster fans on pulverised-fuel fired boilers (section 14.2.2). Similar conditions may also be experienced with the induced-draught fans of solid-fuel fired boilers where high-efficiency dust removal equipment is not mandatory (section 9.4.3.1), for example boilers fired with bagasse (section 15.1.1.1).

The characteristic curves of the radial-bladed fan are illustrated by Fig. 7.5, these indicate that, as the volume increases, the pressure rises to a maximum before falling to zero at maximum flow. The maximum efficiency occurs slightly to the right of the point of maximum pressure.

In the case of the forward- and radial-bladed fans the power consumed increases with increasing volume to a maximum at maximum volume; care must therefore be taken when establishing the power capability of the driver to ensure that it will meet all anticipated requirements.

### 7.2.3 Backward-bladed fans

Backward-bladed fans have a higher efficiency than other types and produce a pressure/volume characteristic similar in shape to that of the forward-bladed fan, as illustrated by Fig. 7.6. Maximum efficiency occurs on the declining section of the characteristic, where the design point is usually selected. The efficiency curve is relatively flat, giving optimum efficiency over a wider range of output than do other blade types.

The power curve reaches a maximum before falling off as the volume increases to the maximum, the driver is therefore less likely to be overloaded than those with forward- or radial-bladed fans should errors occur in predicting the performance.

For a given discharge pressure, backward-bladed fans operate at higher speeds than those with forward or radial blades and hence have higher blade stresses and rates of erosion when used with heavily dust-laden gases. They should only be used for induced-draught fans of solid-fuel fired boilers at low speeds, or if preceded by high-efficiency dust collectors. The blades can be coated with layers of hard materials to improve resistance to erosion.

When fitted with aerofoil-shaped blades, the maximum efficiency obtainable can be in excess of 80%. Under no conditions should aerofoil blades be used with erosive gases, if badly worn they can fill with dust and become badly out of balance. Backward-inclined blades tend to be less noisy than other types.

### 7.2.4 Fan performance

A typical combination of fan characteristic and system resistance curve is illustrated by Fig. 7.7. The gas-side pressure drop (draught loss) across a boiler is substantially proportional to (flow)$^2$, giving a typical square law curve for system resistance. The fan design point is given by the intersection between the system resistance curve and the pressure/volume curve of the fan selected for the application.

Fan discharge conditions are normally specified in terms of flow at a given static pressure and temperature, for example:

5 m$^3$ s$^{-1}$ at 25 mbar and 50 °C.

The fan characteristic can be plotted on the basis of standard conditions (STP). The manufacturers' design data are on this basis but most will supply the user with characteristics based upon the actual conditions under which the fan is to be used.

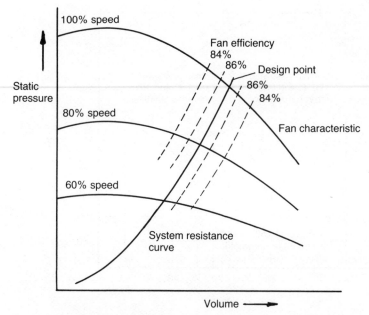

**7.7**   Effect on the characteristic curve of variation of fan speed

There are a number of simple laws that are useful when applying a fan to conditions other than those shown on its characteristic.[2]

1. Volume is proportional to fan speed                        $V \alpha n$
2. Pressure is proportional to (fan speed)$^2$               $P \alpha n^2$
3. Power is proportional to (fan speed)$^3$                   $kW \alpha n^3$
4. Pressure is proportional to gas density                   $P \alpha \rho$
5. Power is proportional to gas density                       $kW \alpha \rho$
6. Volume delivered remains unchanged with density.

Their application is explained simply in relevant publications.[3] Figure 7.8 illustrates how fan characteristics change with density or barometric

Pressure generated and power required are proportional to gas density at a given volume

$$\therefore P_2 = P_1 \times \frac{\text{temperature (K) at } P_1}{\text{temperature (K) at } P_2} \times \frac{\text{Barometric pressure at } P_2}{\text{Barometric pressure at } P_1}$$

$$P_2 = P_1 \times \frac{(15 + 273)}{(150 + 273)} \times \frac{880}{1015} = 0.59 P_1$$

from Fig. 7.8, $P_1 = 45$ mbar

$$\therefore P_2 = 45 \times 0.59 = 26.55 \text{ mbar}$$

similarly $kW_2 = kW_1 \times 0.59$

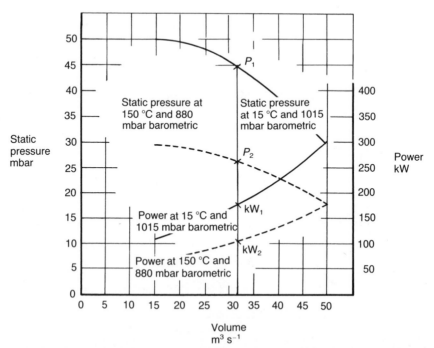

7.8   Example of the application of fan laws to characteristic curves

pressure and temperature. The effect of speed variation is illustrated by Fig. 7.7. In general, a wide fan produces a larger volume than does a narrow one and a large-diameter impellor produces a higher pressure than does a small-diameter one running at the same speed.

On occasions, fan selections are such that, to achieve the static pressure required, it may be necessary to convert some of the kinetic energy to pressure by reducing the gas velocity. In such instances the fan supplier will specify that the fan discharge must be fitted with a divergent duct (evasé) of specified dimensions. This is called 'static pressure regain'. It is therefore essential that the conditions required at the fan discharge are clearly defined and understood.

### 7.2.5 Output control

The simplest and cheapest method of controlling the volume discharged by a fan is by means of a damper installed in the suction or discharge duct to dissipate the pressure that the fan generates in excess of that required to meet plant requirements. Because of their method of operation, dampers are inefficient and tend to have a limited control range, e.g. a turndown capability of perhaps 5 to 1 unless they are very carefully designed. They also tend to disturb the air flow, as mentioned in section 4.1.4. When used with noxious gases, dampers should be fitted in the suction duct of the fan where they will operate under a suction and not create the hazard of gas leakage to the atmosphere through bearings, etc. In these circumstances, their position and design need to be carefully selected if the fan characteristics are not to be affected.

Vanes positioned in the fan inlet are a more efficient means of control than are dampers. They alter the angle at which the gases enter the fan and produce a series of characteristics in the manner illustrated by Fig. 7.9 which all give the same static pressure at zero flow. Fan speed control is the ideal means of fan output control – the characteristic can be continuously altered to suit the system resistance curve such that the fan is always operating close to its point of maximum efficiency, see Fig. 7.7. The speed can be varied by means of variable-speed motors, turbines, or by hydraulic couplings. Variable-speed fans also have the advantage that in an erosive atmosphere they will be running at the minimum speed for a given output, which will minimise fan wear and noise. Variable-speed fans are now being more frequently specified even for the smaller boiler. Fans with variable-speed drives can be less responsive than vanes and dampers to changes of requirements, due to the inertia of the fan. This can be a significant factor if a boiler plant is subjected to frequent and rapid load changes.

Where a boiler is capable of being fired by two fuels giving widely differing gas flows, and these are fired for significant periods

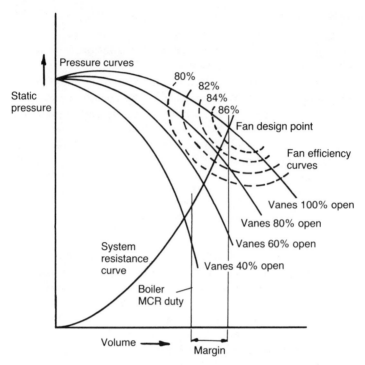

**7.9**  Effect of inlet vane opening upon fan characteristic curve

independently, some benefit can be obtained in power consumption and
turndown capability by having a two-speed drive. This would normally be
used in conjunction with vanes. The two speeds are selected such that full
load on the better quality fuel can be achieved at low speed, hence running
the fan at optimum efficiency. An example where a case can be made for
two-speed fans is the induced-draught fan of a boiler fired with oil and
blast furnace gas, the low speed being used on oil.

### 7.2.6  Design criteria

As fan power is proportional to the density of the fluid being handled, the
maximum power required from the fan drive will be established by the
lowest temperature at which the fan will operate. It is essential therefore
that in the case of forced-draught fans the minimum ambient temperature
at which the fan will operate is specified to the supplier. To ensure that a
sufficient mass of air is supplied under all conditions, it is also essential that
the volume of air required at maximum ambient temperature is specified.

With induced-draught fans, it is usual for them to be started against
closed dampers and, as they are unlikely to produce a high volume at low

temperature, the size of the drive is determined for the maximum duty under normal operation.

Fans must also be designed for the atmospheric pressure at which they will operate. A boiler operating at a high altitude will have a higher system resistance than one at sea level for a given gas mass flow and the fans will also have to handle a higher volume due to the lower density. Both boiler and fan design engineers must therefore be made aware of the altitude of the site above sea level. When applying fan curves care must be taken to ensure that the atmospheric pressure at which they are produced is known. Conversely, if an installation has been designed for operation at high altitude and, for some reason, it is diverted to a low-altitude site, either fan modifications or larger motors are necessary, otherwise the fan motors could be overloaded.

Fan manufacturers will either produce characteristic curves peculiar to the design conditions for a specific project, or they may give the characteristic of a standard fan at standard temperature and pressure. In the latter case care should be taken to ensure that the fan will give the desired conditions at the operating pressure and temperature. This can be checked by applying the fan laws, as illustrated by Fig. 7.8.

To allow for design and operational uncertainties, boiler fouling, wear, etc., margins are added to the boiler requirements at maximum load to determine the fan design requirements. The magnitude of the margins selected will depend, among other things, upon the consistency of the fuel and its fouling properties. Typical margins determined by experience are 15% added to the volume and 32% to the pressure, as the pressure loss varies as the square of the volume, i.e. $(1.15)^2 = 1.32$. The power requirements of the driver are determined for the fan design conditions, or the maximum power given by the characteristic over the operating range, in the case of self-limiting power characteristic (section 7.2.3), plus a further margin of about 10% to give the installed driver excess power capability. The power absorbed at the fan shaft can be calculated at any point on the fan characteristic by use of the following equation:

$$\frac{\text{volume in m}^3 \text{ s}^{-1} \times \text{pressure generated in mbar}}{10 \times \text{fan efficiency}} = \text{kW}$$

Excessive margins should not be added, particularly if the boiler will operate for a significant proportion of its life at reduced load, where the fans will operate at reduced efficiency. The margins applied to a forced-draught fan can normally be less than those applied to an induced-draught fan since the air supply side of a boiler is rarely subjected to fouling except where regenerative airheaters are used or where the boiler is pressurised, i.e. operates under forced-draught only. A forced-draught fan will not suffer from blade erosion or deposition.

When estimating the performance required from a fan, the total system resistance up to the fan suction flange and from the discharge flange onwards must be determined, making allowance for all boiler components contributing to the resistance and for all interconnecting ductwork.

To achieve the desired discharge pressure, the fan manufacturer may specify an evasé or increase in duct cross-sectional area on the discharge of the fan. This reduces the velocity of the gases and converts some of the kinetic energy into pressure. This must be allowed for in the ducting design.

To give improved plant availability with larger water-tube boilers and perhaps improved plant layout and gas and airflow distribution, two fans may be run in parallel, each handling 50% of the required volume at the desired pressure. With this arrangement it is possible to achieve about 70% of the full boiler output when one fan only is in operation and still maintain the margins on volume and pressure. This is indicated by Fig. 7.10 and is possible because the boiler system resistance varies according to the square law. If operation with one fan only is anticipated, tightly closing dampers

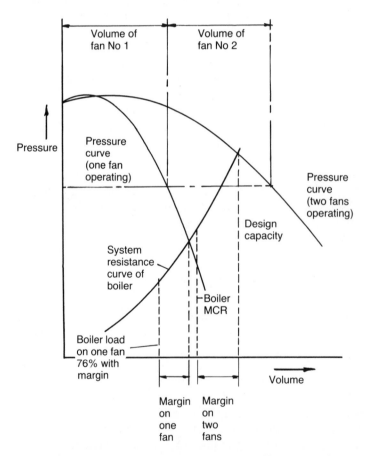

**7.10**   Example of operation with twin fans installed in parallel

are necessary to avoid recirculation through the non-operating fan and hence loss of performance of the operating one.

Centrifugal fans can have single or double open inlets or single or double box or ducted inlets. The choice can often be influenced by the plant layout. Forced-draught and secondary-air fans may have open inlets drawing air from the immediate vicinity of the fan. Care must be taken to ensure that the air flow into the fan is not impeded, for example by being placed too close to a wall. Some form of protective mesh must be fitted over the inlet to prevent entry of any solid objects, or injury to personnel. Silencers are now often fitted in this position and incorporate these functions. They impose a resistance for which allowance must be made on the fan duty. Induced-draught fans have box or ducted inlets for connecting to the boiler. Care must be taken to ensure that if suction dampers are fitted these do not influence the flow of gas into the fan impellor in such a way as to impair the performance. When dampers are fitted to the suction ducting these should be of the multiple bladed contra-rotating type to give minimum disturbance to the gas flow.

The gas or air flow from the centrifugal fan discharge will have a non-uniform velocity profile, with the centrifugal forces tending to throw the air to the outside of the discharge duct. It is therefore particularly important to allow sufficient length of ductwork or make special provisions to even out the flow before it passes through a heat exchanger or to an oil or gas burner. This is particularly important where multiple burners are used and supplied with air from a common duct or windbox as, illustrated by Fig. 1.11, to ensure that each burner receives the correct amount of combustion air. Fan impellors can be mounted in the following manners:

1. on the end of the shaft of the driving motor;
2. overhung on the end of a separate shaft mounted in two bearings positioned between the impellor and the driver; and
3. on the centre of a shaft mounted between two bearings.

Only small fans used for handling clean cold gases should be mounted on the motor shaft, and only then if the motor supplier is in agreement, to ensure that the shaft and bearings have the capability of accepting the loads imposed on them. Large fans should be mounted between bearings for maximum stability.

## 7.3  Chimneys

The purpose of a chimney connected to a modern boiler plant is to discharge the flue gases for dispersal at such a height that they will not create a nuisance to the surrounding district. In the UK the minimum height for a given application and environment is subject to the control of the Chief Environmental Health Officer or the Industrial Air Pollution Inspectorate using the Clean Air Act Memorandum *Chimney Heights*[4] as a guide to determining the minimum height required.

Early boilers operated under natural draught conditions, using the suction created by the hot gas in the chimney to overcome the boiler system resistance. High, large-diameter chimneys were necessary together with high flue gas temperatures. On modern balanced-draught boilers, chimneys are preferably designed to give a slight suction at the base to avoid pressurisation of the gas ducts between the induced-draught fan and the chimney, hence eliminating leakage of flue gases to the atmosphere at ground level. A similar situation can be created for a boiler designed for pressurised operation. The draught available from a chimney and the friction losses through it can be estimated using suitable published data.[5,6]

Having established the required height of the chimney for a given site the materials of construction can be considered. The most common materials for fired boiler applications are steel, concrete and brick, with glass reinforced plastic receiving some consideration, although it is subject to temperature limitations. Chimneys 60–75 m high are likely to be most cheaply made of steel, and those above 75 m high are most cheaply made of concrete.[7] As a consequence, steel tends to be the most common material for industrial boiler applications.

The life expectancy for steel can be 10 years or less for an unlined chimney and up to 30 years for a lined chimney.[8] For brick the life can be as high as 100 years, hence the number of old brick chimneys still in use. Concrete and brick chimneys require larger foundations than do steel ones, due to their significantly greater weight. They can, however, be more readily shaped to suit architectural requirements. Slender steel chimneys can be guyed to give them structural stability, providing suitable anchor points are available.

Chimneys can be single-flue or multiple-flue in a steel or concrete outer structure, or a number of chimneys can be carried by an open steel or concrete structure to give the necessary stability. Multiple-flue arrangements formed by dividing the chimney structure can be subjected to the effects of temperature differentials which set up stresses and bending of the structure. Multiple-flue arrangements, with one flue for each boiler enable minimum velocities to be more easily maintained to avoid inversion and downwash. Inversion is the flow of cold air down the inside of the chimney causing cold spots at the top of the structure. Downwash is the flow of gases down the lee side of the chimney. Both of these can be prevented by maintaining minimum recommended velocities given in the Clean Air Act Memorandum *Chimney Heights* and by other methods.[6]

Chimneys will be subjected to low-temperature corrosion and the formation of acid smuts when used in conjunction with boilers firing sulphur-bearing fuels if the temperature of the material in contact with the gases is allowed to fall below the acid dew point. Care therefore has to be taken with the selection of the final gas temperature from the heat-recovery equipment and with the design of the chimney if this is to be prevented. A chimney can be considered in the same manner as a heat-exchanger tube to determine the structure temperature. For an unprotected steel chimney it will be about midway between that of the gas and that of the surrounding atmosphere.[7] The most susceptible section of the chimney will be at the top where the gas temperature will be the lowest due to the heat lost from the gas as it flows up the chimney. The minimum gas flow condition will give the worst operating situation unless action is taken with the heat-recovery equipment to raise the final gas temperature at low boiler load.

Various methods are available to reduce the possibility of corrosion on a chimney, the most common being the following.

1.  Line the chimney with impermeable acid-resistant materials. These give adequate protection as long as they continue to be impervious to the acid or gases.
2.  Fit internal steel liners, with insulating materials between them and the chimney structure.[9] This prevents the gases coming into contact with the main chimney structure, and the liners can be replaced more readily than can the chimney should corrosion occur.

Where a boiler operates intermittently with, say, an overnight shutdown during which the chimney will cool down, acid condensation is likely to occur during start-up with a buildup of carbon and ash sticking to the internal face. As the temperature rises these deposits dry out and flake away from the chimney, and are discharged as acid smuts. Prevention of acid smuts is difficult with intermittently operating plant. The only sure method of prevention is to maintain the chimney permanently at the desired minimum temperature or to use a neutralising agent[10,11] (section 10.4.2.2).

## 7.4   Heat recovery

As the purpose of a boiler is to transfer heat from a fuel into a working fluid such as water, it is essential that the temperature of the products of combustion be reduced to a minimum in order to obtain the maximum recovery of this heat. The gases leaving the evaporative convection surfaces of a boiler will be at a temperature above the saturation temperature of the water in the boiler, and therefore are likely to be at a temperature of 200 °C upwards depending on the operating pressure. To achieve lower gas temperatures than this necessitates the use of a cooling fluid at a temperature some 50 °C below the final desired gas temperature. The two readily available, relatively cool fluids associated with a boiler are feedwater and combustion air. It becomes natural, therefore, to use one or both of these fluids in heat-recovery equipment to achieve a lower final gas temperature and hence a greater thermal efficiency of the boiler plant as a whole.

Heat exchangers in which feedwater is used to recover heat from the products of combustion are called economisers, the name probably arising from their original introduction to 'economise' on the use of fuel. Heat exchangers in which combustion air is used to recover heat from the products of combustion are called 'airheaters' or 'air preheaters'. To differentiate them from other forms of airheaters in which steam or other hot fluid is used to recover heat or preheat combustion air, they may be called 'gas airheaters'.

On boilers having a significant amount of evaporative convection surface, for example two-drum boilers with a convection bank, as illustrated by Fig. 1.11, operating at pressures of 45 bar and below, it is possible that either economisers or airheaters may be used as the only means of heat recovery. At higher pressures and with single-drum boilers it may be necessary to use a combination of both as, illustrated by Fig. 13.3, to achieve the desired efficiency.

Economisers are not often used in connection with firetube boilers, largely because these operate at relatively low pressures and the exit temperatures are quite low (about 220 °C). In recent years, however, economisers have been increasingly used with gas-fired shell boilers, giving improvements of efficiency of around 4%.[12] The extended use of neutralising substances (section 10.4.2.2) could widen their use to oil- and coal-fired shell boilers, giving a similar enhancement of efficiency. Airheaters are not now used with firetube boilers, although in the 1950s the 'Super Lancashire' boiler incorporated airheaters as standard. These were coal-fired and showed promise, but the increased use of heavy, i.e. less-refined or high-viscosity fuel oils, along with the risks of corrosion associated with acid condensation at low temperatures (Ch. 10) has now made them obsolete. The current change in fuel utilisation patterns could now revive interest both in economisers and in airheaters for use with firetube boilers. Used together, overall efficiencies approaching 90% based on the gross calorific value of the fuel would be feasible.

### 7.4.1 Economisers

#### 7.4.1.1 Economiser types

These can be constructed from plain or gilled cast-iron tubes, steel tubes with gilled cast-iron sleeves or steel tubes only, the steel tubes being either plain or with extended surfaces, the last being more commonly referred to as gilled or finned tubing in this application. (See Fig. 7.11 for examples.)

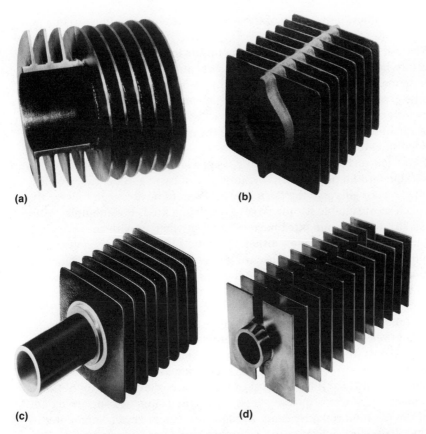

**7.11**  Types of gilled tubes used for economisers: (*a*) helical steel fin; (*b*) cast-iron tube with integral fin; (*c*) cast-iron finned sleeve on steel tube; and (*d*) welded steel gill or fin on steel tube
(From Senior Green Limited)

Early designs of economisers used plain cast-iron tubes with external scrapers for cleaning on the gas side. Later, cast-iron tubes having external cast fins were introduced to increase the heat-transfer surface on the gas side and also to give higher average metal temperatures for a given water temperature inside the tubes.

Cast-iron tubes are still used at pressures below 45 bar (design), where they have an economic advantage. They also have an advantage over steel tubes in that they have better resistance to both internal and external corrosion. In the past their use has been controlled by the provisions of British Standards 1712 and 1713 but these have now been discontinued. Economisers constructed from cast iron have flanged bends joining the straight tubes together to form the water paths. These bends can be readily removed to allow internal mechanical cleaning to be used, this being an advantage where the quality of the feedwater leaves something to be desired.

The use of steel tubes enables economisers to be applied to high-pressure boilers, but as they are more susceptible to corrosion the maintenance of correct feedwater conditions is essential. While steel tubes can be arranged with flanged bends to give an easy cleaning facility these are costly at high pressures. Improved water treatment and the introduction of acid cleaning has led to the almost universal use of fully welded steel tubes.

Extended-surface cast-iron tubing, as produced by one manufacturer, has a standard fin or gill pitch for simplicity of casting. Steel tubes with welded fins or gills give the flexibility for them to be arranged to have the fin pitch most suitable for the type of fuel being fired. For example, with a clean gas, fins at 5 mm pitch have been used, whereas with coal firing a pitch of 20 mm may be more usual. With fuels likely to cause severe fouling, such as municipal refuse, or even some coals, plain tubes may be the most satisfactory solution. The fins can take a wide range of forms, with the economiser manufacturers advocating the virtues of their own designs. Extended-surface economisers occupy less volume than do the plain tube type under the same conditions and, as a result, taking into account the cost of the support structure and enclosure, are usually lower in capital cost for a given set of operating conditions.

Economisers can be 'steaming', that is acting as an evaporator, but these can only be justified if (a) the cost of the surface is less than that of the equivalent evaporator surface, or (b) sufficient evaporator surface to perform the required duty is not available or cannot be accommodated within the boiler enclosure, for example when a single-drum boiler with no convection evaporative surface is being used (Fig. 2.5). Steaming economisers should be avoided at all costs where the quality of the feedwater is suspect because of the risk of solids being deposited inside the tubes within the steaming section. Under no circumstances should cast-iron tubes be used for steaming economisers. BS1712 specified that the water outlet temperature of a cast-iron economiser should not be closer than 40 °F (22 °C) to the saturation temperature of the water in the boiler.

### 7.4.1.2 Corrosion protection

With economisers the tube metal temperature is to all intents and purposes the same as that of the water in the tubes, the heat-transfer coefficient between the water and tube being very high compared with that on the gas side (Ch. 5).

When burning a fuel containing sulphur, care must be taken to ensure that the tube metal temperature is above the acid dew point of the gases if low-temperature corrosion is to be prevented (Ch. 10). Flyash (section 9.4.3) and other particles are also likely to adhere to the sticky acid film, causing blockage of the gas paths. Control of the metal temperature can be performed by maintaining the temperature of the feedwater entering the tubes at the desired value by using feedwater heaters or hot-water recirculation to the economiser inlet. When a de-aerator is included in the feedwater system it may be possible to operate it at a pressure such that the temperature of the water leaving the de-aerator is the desired economiser water inlet temperature. Indirect feedwater heaters can be utilised, with steam or hot water acting as the heating medium. An example of such a system would be passing the feedwater through tubes arranged within the boiler in such a way that the boiler water would be used as the heating medium.

Recirculation of saturated water from the boiler or hot water from the economiser outlet and mixing this with the incoming feedwater requires a pump to give the necessary differential head to overcome the system resistance. Also required is some means of controlling the recirculated water flow to maintain the desired temperature, e.g. a valve controlled by a signal from the mixed water temperature, (see Fig. 7.12). Recirculation

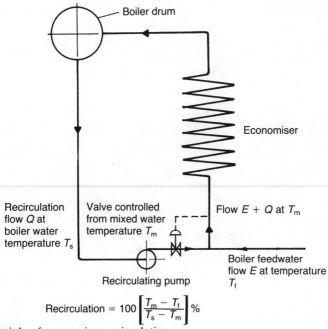

Boiler drum

Economiser

Recirculation flow $Q$ at boiler water temperature $T_s$

Valve controlled from mixed water temperature $T_m$

Flow $E + Q$ at $T_m$

Boiler feedwater flow $E$ at temperature $T_f$

Recirculating pump

$$\text{Recirculation} = 100 \left[ \frac{T_m - T_f}{T_s - T_m} \right] \%$$

**7.12.** Principle of economiser recirculation

of water from the economiser outlet is only practical if the water outlet temperature is significantly higher than the desired mixed inlet temperature. As an approximate guide, the amount of recirculation required from the economiser outlet as a percentage of boiler output is given by the relationship

Recirculation =

$$100 \quad \frac{(\text{recirculated temperature} - \text{feedwater temperature})}{(\text{economiser water outlet temperature} - \text{recirculated temperature})}$$

The performance of the economiser, particularly at low loads, may not be sufficient to enable the desired effect to be achieved.

While recirculation of water from the boiler will achieve the desired effect, it may cause deposition of solids in the economiser tubes when trisodium phosphate is used for dosing the feedwater. Indirect feedwater heating is therefore a preferred method of controlling the economiser metal temperature to prevent the occurrence of low-temperature corrosion. The use of neutralising additives, such as magnesium hydroxide to the products of combustion is a more recent approach and shows much promise, particularly on existing plant where operating conditions have changed.

It must be appreciated that when using extended surface tubing, the metal temperatures at the root of the fin, i.e. nearest to the water in the tube, will be at water temperature. It is therefore this point which controls any action taken to protect the tubes against corrosion, even though most of the surface within the fins is at a higher temperature. It must be appreciated that elevation of the feed water temperature will necessitate increased economiser heating surface for a given heat transfer because of the reduced temperature difference (see section 5.6).

### 7.4.2   Airheaters

These can be of the recuperative or regenerative types. In the recuperative design, the gas and air are separated by the heat-transfer surface. Heat is transferred from the hot gases on the one side through this heat-transfer surface to the air on the other side. In the regenerative design, the heat is transferred indirectly by first heating a storage medium with the hot gas and then bringing this storage medium into contact with the cooler air to transfer the heat to the air.

Airheaters are non-pressure parts and may be preferred to economisers where the feedwater quality is such that economisers would be subjected to corrosion and where suitably qualified labour may not be readily available to carry out repairs.

Recuperative airheaters used with boilers are of the tubular or plate type.

#### 7.4.2.1   Tubular airheaters

A tubular airheater can be constructed with plain tubes, the ends of which are attached to tube plates, or with cast-iron tubes having extended surfaces on both the air and gas sides. The gas can be arranged to pass

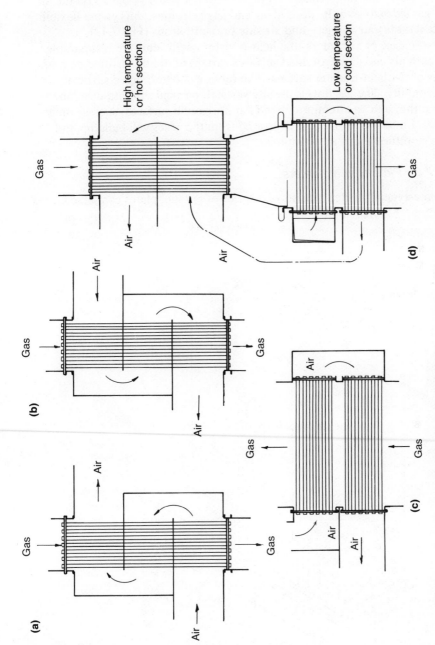

7.13   Examples of air and gas flow arrangements used with tubular air heaters:
(*a*) counterflow; (*b*) cocurrent flow; (*c*) gas across tubes, air through tubes,
counterflow; and (*d*) cold end in cocurrent flow.

either through or across the outside of the tubes, as desired for a particular application. The tubes can be arranged with a number of passes on the air and gas sides to give the desired air and gas velocities, thus giving flexibility with plant layout and gas- and air-side pressure drops (Fig. 7.13).

In the case of airheaters, the heat-transfer coefficients are comparable for both air and gas. Extended surfaces can therefore be justified on both sides of the heat-transfer surface. The tubes can be arranged vertically or horizontally. The gas flow is usually vertical, particularly with dust laden gases; this is in an attempt to avoid flat horizontal surfaces, such as tube plates, on to which any dust deposited from the gases will build up and may be difficult to remove on-load.

### 7.4.2.2  Plate-type airheaters

A plate-type airheater consists of a series of parallel plates connected in

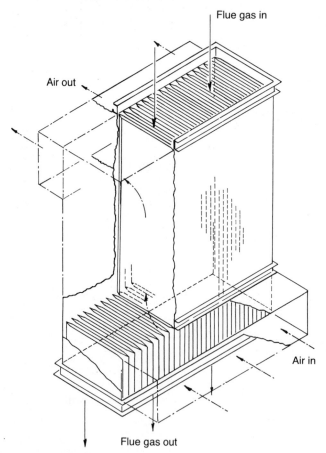

**7.14**  Plate-type air heater

such a way that the gas and air flow through alternate spaces between the plates. The gas flow is usually a straight path with the air flow changing direction at the inlet and outlet to simplify the interconnecting air and gas ducting (Fig. 7.14). The plates of such an airheater can be either plain or have extended surfaces on both sides to enhance heat-transfer rates. The plate temperature is approximately midway between the gas and air temperatures adjacent to it. Extended-surface airheaters will occupy a smaller volume than those constructed from plain surfaces.

### 7.4.2.3  *Regenerative airheaters*

Regenerative airheaters are of two basic types, having either rotating or stationary heat-storage elements. These elements consist of a cylindrical bundle of corrugated or undulating metal plates, as illustrated by Fig. 7.15, arranged parallel to the axis of rotation. The undulations create the spaces between the plates through which the gas and air flow parallel to the axis of rotation. With the rotating element (Ljungstrom) type, illustrated by Fig. 7.16, the heat-storage elements rotate at up to 3 rev min$^{-1}$, each part passing alternately through the gas duct where it becomes heated and then through the air duct where it is cooled down, thus heating the air in the process.

With the stationary element (Rothemühle) type, illustrated by Fig. 7.17, the gas and air ducts rotate around the axis of the elements to give the desired alternating air and gas flows over the heat-storage medium.

With each type of rotary heater, sealing between the moving and stationary parts is important to prevent air leakage from the high-pressure air side to the low-pressure gas side, which would thus bypass the storage medium and reduce the thermal performance. Various forms of self-adjusting seals are used, but maintenance of gas-tight seals and hence a low level of leakage over the period of operation between maintenance outages is difficult.

Rotation can be arranged about the vertical or the horizontal axis to suit the plant layouts. Solid-fuel fired boilers normally have a vertical spindle arrangement to simplify cleaning. The cleaner fuels of gas and oil fired units are more suitable for a horizontal-spindle arrangement, which can result in a much more compact ducting layout.

Regenerative airheaters are very compact and occupy much less space than the equivalent tubular type and are almost universally used on the large utility boilers of 120 MW electric output and above. The smaller sizes are shop assembled, or packaged, hence reducing site construction work. The effective metal temperature for a regeneration airheater is difficult to determine as it is changing continually while the elements are heated and cooled. The sum of the air inlet and gas outlet temperatures is used as a guide to the operating metal temperatures. Manufacturers can now readily tabulate metal temperature by the use of computer programs.

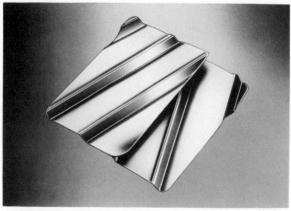

Flat notched crossed (FNC) heating surface elements for application in clean gas systems where little fouling is anticipated

Double undulated (DU) heating surface elements for application where only moderate fouling is anticipated

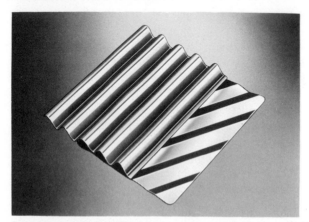

Corrugated undulated (CU) heating surface elements for application where fouling is anticipated

**7.15**  Types of rotary airheater heating surface elements
(From Howden Group plc)

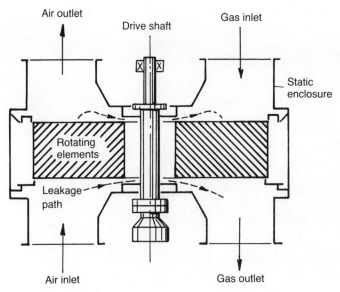

Air outlet

Drive shaft

Gas inlet

Static enclosure

Rotating elements

Leakage path

Air inlet

Gas outlet

**7.16** Rotating heating surface (Ljungstrom) airheater
(After Howden Group plc)

### 7.4.2.4 Corrosion protection

As with economisers, airheater surfaces can be subject to low-temperature corrosion if the temperature of those surfaces is allowed to fall below the acid dew point of the gases. As airheaters are non-pressure parts it is, however, possible to use materials having a greater resistance to acid corrosion than steel tubes. Glass tubes have been used successfully on tubular heaters[13] and ceramic or enamelled elements in the rotary type.

Tubes with various coatings on the gas side have been used, but unless the protective coat is completely impervious to acid the base material will still be affected by corrosion. There are, of course, cost penalties in using these alternatives and, where possible, it is preferable to achieve suitable metal temperatures by careful design of the heating surface configuration, or use of a neutralant.

Methods by which maximum airheater heat transfer surface temperatures can be achieved are given in the following sections.

(a) Arrange the gas and air in cocurrent flow where possible, particularly within the cold section of the heater, as illustrated by Fig. 7.13(*d*).

(b) Install steam- or hot-water-heated preheaters in the air duct before the gas airheater to raise the temperature of the air entering the gas airheater. The steam or other fluid used for heating can contain excess heat available from a process line or be pass-out steam from a turbine. High-pressure steam from the boiler can be used in the

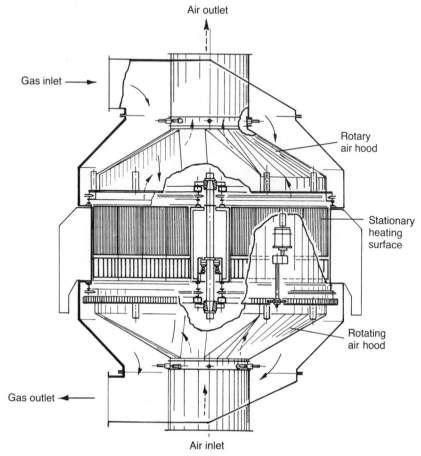

Air outlet

Gas inlet ⟶

Rotary
air hood

Stationary
heating
surface

Rotating
air hood

Gas outlet ⟵

Air inlet

**7.17**  Stationary heating surface (Rothemühle) airheater
(After Davidson & Company Limited)

absence of any other suitable source of heat, the disadvantages of this
method being the feed-pump power absorbed to handle the extra
water flow, the heat loss from the piping system and the problem of
handling the condensate from the preheater. The heating medium
used must of course be at a temperature above the required air outlet
temperature.

(c)  Recirculate hot air from the airheater air outlet to the air inlet via the
forced-draught fan in sufficient quantity to raise the mixed air
temperature to the desired value. This method increases the air-flow
area required for a given velocity, or the air-side pressure loss, and
also the forced-draught fan duty. Recirculation can be advantageous
at reduced loads and low ambient air temperature without having
significant effect on the heater design.

(d) With a tubular heater, arrange the gas flow across the tubes and the air flow through them, as illustrated by Fig. 7.13(c), to give a higher heat-transfer rate on the gas side.

(e) At low loads and at start-up, bypass some of the air around all or part of the heating surface, hence raising the temperature of the air in the heater and reducing the air-side heat-transfer rate.

(f) Use a neutralising additive to the products of combustion (section 10.4.2.2).

The above courses of action can be used singly or in combination, but if used at full load will all result in an increased heating surface to achieve the desired heat transfer. When fuel costs are high it may be justifiable economically to incorporate a disposable section of heating surface at the cold section, which can be readily replaced at planned regular intervals. Accurate prediction of the life of such a section of heating surface is virtually impossible, due to the large number of factors involved – experience is the best guide.

While the above comments refer to the heat-transfer surfaces, attention must also be paid to sections such as the tube plates of tubular heaters where both cold air and cold gas come into contact with the material.

### 7.4.2.5 Performance of airheaters

The performance that can be obtained from an airheater is limited by the following factors:

1. the maximum temperature that the combustion equipment can withstand, for example the air temperature to stokers is restricted to 175/200 °C, but for oil burners this can be as high as 400 °C without the use of special materials; and

2. the fact that the air temperature rise is greater than the gas temperature drop due to the gas flow always being higher than the air flow. This has a large influence with low-grade fuels such as blast-furnace gas (see Table 7.1).

**7.1** Comparison of gas and oil flows for blast furnace gas and oil

|  | Blast Furnace Gas | Oil |
|---|---|---|
| Gas flow | 1.98 | 1 |
| Air flow | 0.776 | 0.94 |
| $\dfrac{\text{Gas flow}}{\text{Air flow}}$ | 2.55 | 1.07 |
| $\dfrac{\text{Air temperature rise}}{\text{Gas temperature fall}}$ in an airheater | 3.06 | 1.28 |

## 7.5 Sootblowers and boiler cleaning

To ensure that the performance and thermal efficiency of a boiler are maintained it is essential that the heated surfaces are maintained clean. On the gas-swept surfaces, this necessitates removal of material deposited on the tubes from the flue gases. If this is not done the rate of heat transfer from the gases will reduce and the gas temperatures will rise, perhaps aggravating the problem. On most solid-fuel fired boilers and, dependent upon the fuel properties, on some gas- and oil-fired and waste-heat boilers, sootblowers are installed to enable the boiler surfaces to be cleaned while the boiler is operating.

A sootblower is a device that directs a jet of steam or compressed air to blow across tube surfaces in contact with the flue gases with the intention of removing material deposited on the tubes. They can be of the multi-nozzle or multi-jet rotary type or of the retractable type.

A multi-nozzle sootblower consists of a steel tube of 50–64 mm diameter which is inserted through the wall of the boiler and projects into the gas stream, with holes (nozzles) at intervals along it, through which the blowing medium issues. Such a blower as illustrated in Fig. 7.18. The nozzles are positioned to coincide with the spaces between the tubes to enable the steam to blow down the gas passages between the tubes. The blower can be rotated through any angle up to about 280° to cover the greatest amount of heated surfaces. Where it is required to blow around a full 360°, two rows of diametrically opposite nozzles are used and the blower is rotated through 180°. The effective radius of cleaning from the

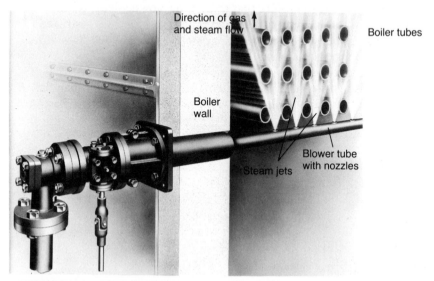

**7.18** Multi-nozzle rotary sootblower
(From Hopkinson Limited)

centreline of tube is about 2 metres. Multi-nozzle blowers which remain in the gas stream can only be used in gas temperatures up to about 1000 °C due to the lack of suitable materials of construction for higher gas temperatures. Their use is therefore mainly restricted to the evaporative convection, economiser and airheater surfaces.

Where gas temperatures exceed those for which fixed blowers are suitable, the retractable type have to be used. These can be either short or long. With the short type the nozzle projects just beyond the boiler wall and can be used to blow either the combustion chamber wall tubes, or the convection heating surfaces on narrow boilers, as illustrated by Fig. 7.19. With retractable blowers the tube is withdrawn from the gas stream when not in use, and there are nozzles only at the end of the tube. When sootblowing is being carried out with a long retractable blower it is rotated and at the same time traverses the gas stream, so covering the full width of the boiler, the steam jet following a helical path. The full cycle includes blowing while the tube traverses back across the boiler and is withdrawn. The steam issues from the jets immediately they enter the gas stream to ensure that the tube is always adequately cooled. Long retractable blowers have opposing nozzles at the end to ensure that the reaction of the steam jets is balanced so as to reduce deflection of the tube.

The tube is available in lengths up to about 15 m as required by the width of the boiler; blowers can be fitted in both sides if required to reduce the length of the tube. On wide boilers allowance has to be made in the layout of the heated surfaces for the deflection of the tube due to its own weight.

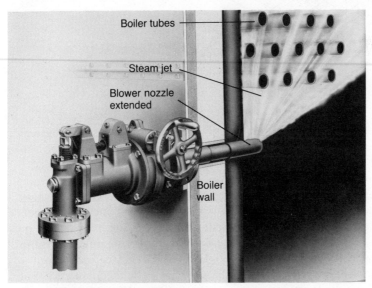

**7.19**   Short retractable sootblower
(From Hopkinson Limited)

Ample space has also to be allowed outside the boiler to enable the tube to be fully withdrawn when not in use. With multiple boiler installations the use of long retractable blowers can dictate the space between boilers. Because the blower traverses the boiler, there are inevitably occasions when the steam jets are directed on to the tubes; to avoid tube erosion by the blowing medium the centreline of the tube should be at least 500 mm from the nearest tubes.

For gilled or finned tube surfaces of the type illustrated in Fig. 7.11 it is preferable for the steam jets to blow down the tube lanes between the fins to give adequate cleaning. Rake type traversing blowers of the configuration illustrated by Fig. 7.20 are used to achieve this.

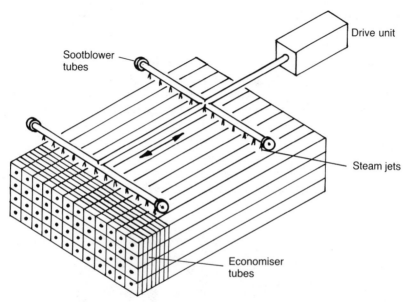

**7.20**   Traversing rake type sootblower

The sootblower manufacturers will give recommendations regarding the positions, numbers and types of blower to use for a given application.

It should be appreciated that sootblowers each use steam at a rate of $2–4$ t $hr^{-1}$, which on small industrial boilers can be a significant percentage of the boiler output for the duration of the blowing sequence. This can be several minutes, depending upon the width of the boiler and the number of blowers fitted. Because of this rate of steam consumption, blowers are only used one at a time to reduce the effect upon the boiler steam production. For this reason also, sootblowing should preferably only be carried out during periods of low load. The steam usage has to be taken into account when designing the feedwater system (see sections 6.10 and 7.1).

Multi-nozzle and short retractable blowers can be manually operated by hand wheels, but the effectiveness of blowing is then dependent upon how the operator manipulates the blower. It is more usual now for these to be electronically operated.

Long retractable blowers are always electronically operated to ensure that the sequence is completed and the blower safely withdrawn from the gas stream. While the sootblowers can be operated locally or remotely, automatic sequence operation is now more usual, with programmable controls enabling the sequence of operation of the various blowers to be varied to suit the cleaning requirements of the boiler.

The use of sootblowers applies mainly to water-tube boilers, where regular on-load cleaning is essential. In most cases firetube boilers are not required to operate for such long continuous periods and sootblowers are not often fitted. The boilers are shut down at weekends, the smoke-box door opened, and the tubes swept by brushes which may be mechanised and equipped with a vacuum extraction device to remove the loosened material. An effective method is to use a 'percussion lance'. This discharges rapid pulses of compressed air down each tube to which the lance is presented. Both brushes and lances are hand-held and are not attached to the boiler.

A more recent method of sootblowing, still to achieve its full potential, is to use intense sound. This is the 'sonic' sootblower which, as with the percussion lance, discharges rapid pulses of compressed air into the cavities in the boiler which it is designed to clean. The frequency of the pulses can be tuned to resonate with the natural frequency of the cavity, so increasing the amplitude of the pulse. In many cases deposits are of a porous structure, gases existing in the pores. A positive compression wave is followed by a negative wave which causes the gases included in the deposit to expand, thus disintegrating the mass.

Sonic sootblowers are now increasingly used on both firetube and water-tube boilers, and in some cases are proving very effective.

# References

1. K.S.B. HANDBOOK, *Pumps*. Klein, Schanzlin & Becker, Germany, 1968.
2. Buffalo Forge Company. *Fan Engineering*. Richard D. Madison, pp. 189–234.
3. *Notes on Fan Engineering*. Davidson & Co. Ltd.
4. 'Memorandum on chimney heights', *Clean Air Act 1956* (3rd edn). HMSO, London.
5. *Chimneys for Industrial Oil Fired Plant*. Shellmex & BP Ltd.
6. Haines, W. *Notes on Chimney Design*. Institution of Plant Engineers.
7. Beaumont, M. 'Design of industrial chimneys', *Works Eng.* 1971 (July).

8. Jackson, K. R. 'Molar chimneys, *J. Inst. Fuel* 1969 (Sept).

9. Ravenscroft, F. P. and Page, H. A. 'Field investigation and design of insulated chimneys', *J. Inst. Fuel* 1966 (Jan.) 22.

10. Lees, B. 'Long-term experience with magnesia additives in large oil-fired boilers' *J. Inst. E.* 1980 (Sept.) 134.

11. Davis, I. *et al.* 'The use of magnesium-based flue gas additives in minimizing acid smut emissions from oil-fired boilers', *J. Inst. E.* 1981 (Mar.) 21.

12. Gray, K. *et al.* 'Economisers for modern boilers', *J. Inst. E.* 1981 (Sept.) 151.

13. Bowen, I. G. 'Glass air heaters for watertube boilers', *J. Inst. Fuel* 1968 (*Mar.*).

# CHAPTER 8

# Instrumentation and controls

## 8.0 Purposes of instrumentation and controls

'Instrumentation and controls' is a widely used collective term covering two distinct aspects of boiler operation.

Instruments are those items attached to a boiler which give an indication to the operator of the conditions that exist within the boiler and thus enable him to ensure that these are within the safety limits and operational parameters for which the boiler was designed. Examples of indicated quantities are operating pressures, temperatures, flows and water level.

Controls are those items which carry out the function of regulating the various quantities indicated by the instruments and which can be arranged, with interlocks, to shut the plant down if any values pass outside the allowable operating range. Control systems can vary in sophistication from local manual operation of the various valves and dampers to a fully computerised system with little manual intervention once the system is programmed and verified. It is worth reflecting on the statement, 'Before you can control you must measure'. This applies to manual as well as to automatic control.

Manual control is tedious and requires continuous watch on all the instruments to ensure that safe conditions exist. It is also necessary to include alarms to alert the operator to the fact that corrective action is required. To control a boiler, the following quantities require to be regulated as applicable to a particular system:

1. the heat input to the boiler to match the required heat output;
2. the fuel/air ratio to maintain optimum combustion conditions (combustion control);
3. in the case of steam boilers the water flow to match the steam flow from the boiler;

4. combustion chamber pressure in the case of balanced-draught boilers to maintain a small negative pressure on the gas side;
5. where high degrees of superheat are generated, the steam temperature may be controlled to protect the superheater, steam pipework, and the device using the steam against overheating; and
6. combustion safety (burner management).

Instruments and controls associated with carrying out actions 1 to 6 may be interconnected in various ways, either to use common signals or to improve the response of the boiler to a change in one of the parameters.

Due to the rapid development of electronic and control systems, particularly the microprocessor, the systems available for the control of boilers are diverse and continually developing. As a consequence, the purpose of this chapter can only be to describe the principles rather than the details of measuring and control devices for boilers. The scope and complexity of automatic controls and instrumentation can vary enormously from the simple on/off schemes as applied to small firetube boilers to the more complex modulating schemes with extensive visual-display and computer data storage facilities used on some of the larger boilers.

While minimum requirements are mandatory[1], the extent of equipment to be included depends very much upon the users' requirements and their interest in retaining reliable records of the various operating parameters. Continuous recording, as opposed to hourly logged readings, will give a record of any very short term changes in operating conditions which may otherwise pass unnoticed. Such records can be useful when analysing the causes of faults that occur and may also enable impending problems to be anticipated and prevented.

## 8.1   Operating media

Except for instruments which are direct reading, such as pressure and water level gauges, and some control systems that use the fluid being controlled to operate them, for example the feedwater control valve illustrated by Fig. 8.1, the systems use either electricity (electronic) or compressed air (pneumatic) or a combination of both for transmission of signals from the various sensing devices through the controllers to the actuators. The choice between electronic or pneumatic devices tends to be influenced by the availability of compressed air, maintenance considerations and overall cost.

Small installations involving firetube boilers are likely to use electronics because the provision of small quantities of compressed air can make pneumatics uneconomical, and dampers and valves are of such a size that they can be easily operated by electric actuators. For large units and

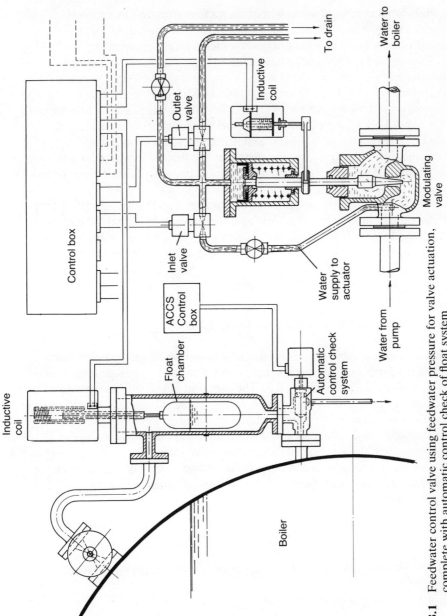

**8.1** Feedwater control valve using feedwater pressure for valve actuation, complete with automatic control check of float system (After Bestobell Mobrey)

multiple boiler installations pneumatic actuation may be favoured since the actuators for large valves and dampers can be less expensive than electric units. Pneumatic controls are also very precise and smooth in action. For installations in remote areas where competent electrical or controls engineers are not available, pneumatics can be favoured, being more readily understood and maintained by mechanical engineers.

Pneumatic systems can be subjected to freezing in cold climates if instrument air is not perfectly dry. The necessity for protection must therefore be considered in any evaluation. With pneumatic systems the air pressure used for signals varies from a low of 0.2 bar (3 p.s.i.) to a high of 1.0 bar (15 p.s.i.) with the full-load working point being equivalent to about 75% range (i.e. 0.8 bar). Due to the pressure drop in the transmission lines, transmission distances are limited to about 80 metres maximum. If greater distances are involved, pneumo-electric transmitters are available.

With electronic systems, due to the voltage drop that occurs along long transmission lines, current is used as a measure of the signal. The signal level used varies from a low of 4 milliamps to a high of 20 milliamps with the full-load working point being equivalent to about 16 milliamps.

## 8.2   Instrument and control diagrams

Instrument and control systems are usually shown diagramatically using standard symbols.[2] A typical simple control loop for controlling furnace gas pressure is shown in Fig. 8.2. The function of the various pieces of equipment is described in section 8.6. All automatic systems should include facilities to select either hand or automatic control to enable the operator to take over in the event of a system failure. This facility is usually incorporated in the pressure indicator/control item PIC.

The remainder of this chapter explains some of the equipment used for the measurement and control of the various parameters associated with fired boilers.

## 8.3   Boiler pressure measurement, indication and relief

### 8.3.1   Local pressure indication

This, along with the corresponding temperature measurement and control with hot-water boilers, is perhaps the most basic function required. First, the display (which must be easily seen and read by the operator) is

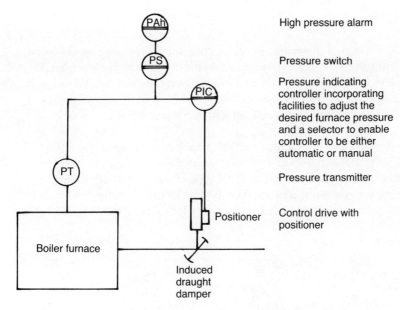

High pressure alarm

Pressure switch

Pressure indicating controller incorporating facilities to adjust the desired furnace pressure and a selector to enable controller to be either automatic or manual

Pressure transmitter

Control drive with positioner

**8.2** Simple diagram for induced-draught fan control

necessary to ensure the safety of the plant, a pressure rising above a clear mark indicating the working pressure on the dial signals that the heat input must be reduced immediately. A falling pressure means that the demand for heat is exceeding the heat input and therefore that the firing rate must be increased. The indicating instrument is the well-known Bourdon gauge which consists of a flat tube bent to a curve. This tends to straighten out as the internal pressure increases and is arranged to drive a pointer over a circular scale.

### 8.3.2 Remote pressure indication

Where the pressure indication needs to be transmitted to a remote indicator, to a recorder or to a microprocessor electronically without having high-pressure lines around the plant, an electrical signal is desired. A separate transducer is used consisting generally of a diaphragm or bellows acting against a calibrated spring. A strain gauge, depending on the change of electrical resistance with deformation, or a contact moving over an electrical resistance is connected to the diaphragm. This translates deflection of the diaphragm or bellows into a change of resistance which is measured on a Wheatstone bridge. This in turn gives an output which refers the pressure to the recorder or microprocessor. These transducers are very compact and can be mounted on the same branch as the pressure indicator.

### 8.3.3 Safety valves

Boilers are designed to withstand certain pressures only, and on no account must be subjected to greater pressures. In most cases the measuring and control devices described suffice to avoid an overpressure condition but it is mandatory, on both steam and hot water boilers, to fit safety valves which lift and relieve the pressure, albeit with much noise and waste of heat.

Safety valves have a long history dating back to 1860 when Dr Denys Papin (FRS) designed a lever-type safety valve for his 'digester'[3], the forerunner of the pressure cooker. This type of valve, and other earlier deadweight types are not now used on boilers as their setting can be easily interfered with. They have been replaced by lockable, spring-loaded types.[4]

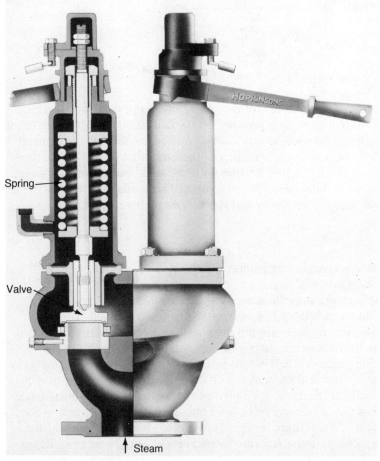

Spring

Valve

↑ Steam

**8.3** Double spring-loaded safety valve
(From Hopkinson Limited)

Even so, they are subject to rigorous maintenance, inspection and certification. A typical example of a spring-loaded valve is illustrated by Fig. 8.3.

## 8.4   Combustion control

This incorporates both the control of the boiler heat input and fuel/air ratio to the combustion chamber. Combustion control systems must ensure that at all times adequate quantities of air are available to meet the fuel requirements, so as to burn the fuel efficiently without smoke and with minimum particulate discharge from the chimney. The main source of signal for the operation of a combustion control system is the steam pressure at the boiler outlet, in the case of steam generators, and the water outlet temperature, in the case of hot-water boilers. Combustion controls therefore also control the boiler pressure as a stage in controlling the heat input.

### 8.4.1   Control schemes

There are three basic control schemes used for regulating multiple variables such as fuel and air flow in a combustion control system. These

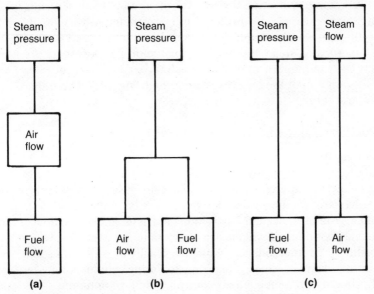

**8.4**   Basic control schemes
(*a*) series control, (*b*) parallel control, and (*c*) series/parallel control

are: series, parallel and series/parallel. With series control a variation of the master control signal, steam pressure, causes a change to take place in the combustion air flow which, in turn, causes a change in fuel flow (Fig. 8.4(a)). With parallel control, a variation of the master control signal adjusts the fuel and air flows simultaneously (Fig. 8.4(b)), and represents a typical positional control system. With series/parallel control, a variation of the master control signal adjusts the fuel flow and, as steam flow is approximately proportional to air flow, variations of steam flow resulting from a change of load are measured and used to adjust the air flow (Fig. 8.4(c)).

## 8.4.2   Types of control system

There are three basic types of automatic combustion control. These are: on/off, positioning and metering.

### 8.4.2.1   On/off systems

On a steam boiler, using an on/off system, the fuel and air are shut off as the steam pressure rises to a preset value. The steam pressure then falls gradually as the demand continues, until it reaches a predetermined low value at which the fuel and air are turned on again.

With hot-water boilers, high and low water temperatures are used as the initiating signals. A typical example of the on/off control is the system used with a gas-fired domestic heating system. This method of control results in a fluctuating steam pressure. Its use tends to be restricted to very small units generating hot water or saturated steam. It cannot be used when generating superheated steam because, during the off periods, there are no gases flowing over the superheater from which the steam can receive its superheat. A variation of the on/off system is 'high/low/off', where there are three control settings instead of two.

### 8.4.2.2   Positioning systems

With positioning systems, the fuel and combustion air controllers (the fuel valve in the case of oil or gas firing, and dampers or fan speed in the case of combustion air) are interconnected mechanically in such a way that for a given fuel valve position the air damper will always be in the same position. Such systems are called 'open-loop' and assume that the flow through the valve or damper will always be the same for a given valve or damper position. The interconnecting linkage usually incorporates some form of cam, the shape of which is determined during commissioning by manual adjustment of the fuel and air controllers to give optimum conditions over the load range of the boiler. Figure 8.5 illustrates such a system.

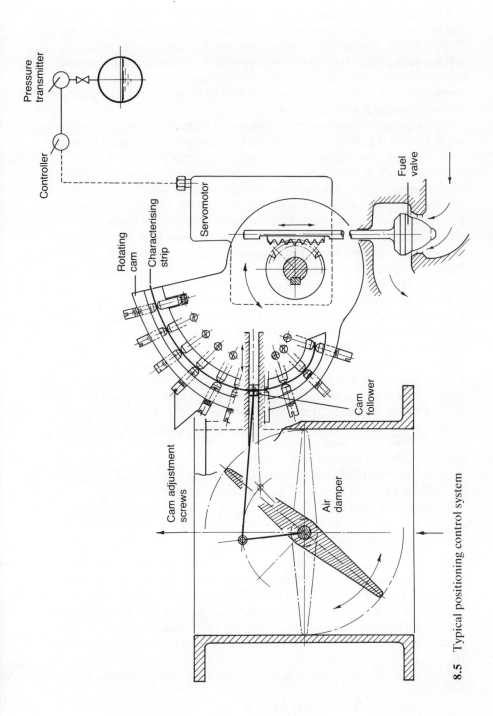

**8.5**   Typical positioning control system

Pressure transmitter

Controller

Servomotor

Fuel valve

Rotating cam

Characterising strip

Cam follower

Cam adjustment screws

Air damper

This method of control is affected by changes of the system resistance on the fuel and air side caused by wear and fouling on the fuel system and fouling on the air/gas side and may need regular adjustment. This is not easy to incorporate so as to maintain optimum conditions. They are also susceptible to hysteresis, resulting from mechanical play or slack and friction in the linkages. Their use is limited to firetube and small water-tube boilers using consistent fuels.

On a typical positioning system applied to firetube boilers the pressure control signal is generated by separate sensors, two of which are generally used. The first is to signal an overpressure condition to the fuel-feed regulator which in turn is linked to the combustion air supply. Should an overpressure condition occur, the firing appliance is shut down, generally accompanied by visual and audible alarms, and needing manual reset. This control is mandatory for automatic boilers.[5,6] The second sends an electrical signal which is proportional to the change of pressure from the set point to a servomotor connected to the fuel regulator and to the air-regulating dampers (or to the fan-motor speed controls). These are thus adjusted to restore the pressure to the set value.

Hot-water boilers employ somewhat similar methods except that maximum water temperature is used for the shut-off signal and often for the modulating signal. The sensor consists of a bellows filled with a wax having a high and stable coefficient of volumetric expansion. As with the pressure sensor, this acts against a spring, any movement being transmitted to a contact which moves over a resistance which forms one arm of a Wheatstone bridge to centre a modulating signal on a microswitch to signal an off condition. Temperature indication can be by a mercury-in-steel thermometer, but the use of indicating thermocouples is now more common. Hot-water boilers also require a pressure gauge to indicate that the head of water necessary to ensure safe operation of the boiler is present, whether this head is produced by gravity or by a pressurising unit. In addition a pressure sensor is necessary to cut off the fuel supply should the head not be present.

### 8.4.2.3   Metering/modulating systems

With metering systems, the fuel and air are regulated by the master signal from the steam pressure, a fall in pressure indicating that an increase in fuel and air inputs is required. The fuel and air flows are measured, the two signals are compared in a ratio controller (feedback) and one of them is adjusted by operating the flow controller until the correct ratio or set point is achieved. The combustion conditions are therefore maintained at the optimum irrespective of any changes that may occur to the system resistance or characteristics of the controller. Such systems are called 'closed-loop'.

The ratio controller is arranged so that the set point can easily be adjusted manually while the boiler is in operation should there be any change in the fuel characteristics and hence in the heat input to the boiler for a given fuel flow signal.

Metering systems do, however, require a flow measuring device in the fuel and air systems which, depending upon the accuracy required, can vary from using the pressure drop across part of the flow system to installing orifice plates and venturis. These last have minimum requirements for upstream and downstream lengths and configurations to give accurate repeatable readings.[7,8] The use of special flow metering devices increases the system resistance and hence power consumption of the auxiliaries. If the temperature of the oil is under close control, the pressure drop across an oil burner tip can be used to measure oil flow, and that across the boiler heating surface or burner air register to give a measure of air flow. This system, however, should be restricted to single-burner installations because the total fuel flow is dependent upon the number of burners in use and making the control system aware of this gives added complications.

### 8.4.2.4  Choice of system

The choice of system depends upon the accuracy with which it is required to control the boiler, and hence on the desire to operate the boiler at optimum conditions at all times. The cost is also a major factor, on/off control being the least expensive and metering control the most expensive. For the fine tuning of combustion controls, flue gas sampling for oxygen and/or products of combustion[9] is incorporated, and a signal is fed back to the control system. The primary signal for oxygen control, or trim, is obtained from a zirconium oxide probe inserted in the flue gas stream. The electrical resistivity of the probe varies with the vapour pressure of the oxygen present and generates the signal for feeding back to the controller.

There is a tendency for the excess air/oxygen content to increase at low loads to ensure that adequate fuel and air mixing is achieved with the lower air and fuel velocities that occur under reduced load conditions. It must therefore be possible to characterise the oxygen correction signal. To do this, a secondary signal is taken from the steam flow transmitter to indicate the actual boiler output.

With coal firing on grates, oxygen trim has as yet to be more fully developed, since, for various reasons, incompletely burned coal can be discharged from the grate even when excess oxygen is present. A signal to reduce the air under such conditions will only worsen the effect. A measurement of unburned coal discharged is therefore needed to supplement the oxygen measurement. No instrument for this last purpose has yet been developed commercially.

For larger plant, the oxygen measurement can be supplemented by carbon monoxide and hydrocarbon measurement, control over which not only maximises efficiency but reduces stack emissions.

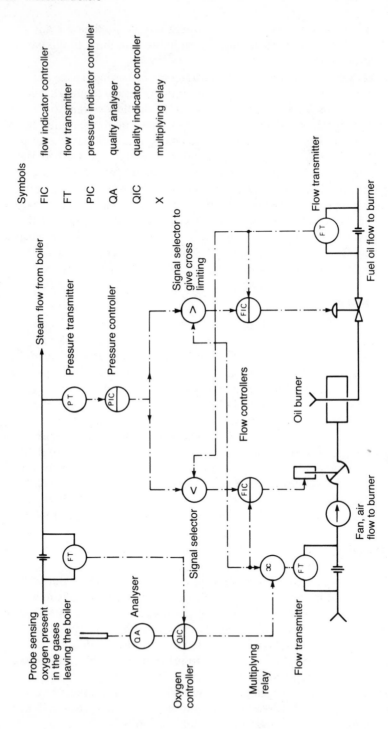

**Symbols**

FIC    flow indicator controller

FT    flow transmitter

PIC    pressure indicator controller

QA    quality analyser

QIC    quality indicator controller

X    multiplying relay

Steam flow from boiler

Pressure transmitter

Pressure controller

Signal selector to give cross limiting

Flow transmitter

Fuel oil flow to burner

Oil burner

Flow controllers

Probe sensing oxygen present in the gases leaving the boiler

Analyser

Signal selector

Oxygen controller

Multiplying relay

Flow transmitter

Fan, air flow to burner

**8.6**    Control diagram for oil-fired water-tube boiler

Whatever methods of control are used, it is necessary that at least the oxygen or carbon dioxide content of the gases be measured and displayed to inform the operator about the combustion conditions.

With gas and oil firing the response of the fuel control valve is usually more rapid than that of the air flow controller. If there is an increase in steam demand from the boiler the fuel input increases more rapidly than does the air input, possibly resulting in the production and emission of black smoke, particularly if low excess air operation is being carried out. To prevent this occurrence, 'lead-lag', 'cross limiting' or 'air lead' is included, which prevents the oil flow from increasing more rapidly than that of the air during an increase of load and prevents the air flow from reducing more rapidly than that of the fuel on reduction of load. Figure 8.6 illustrates a composite combustion control diagram which incorporates oxygen trim and cross limiting.

With stoker firing, and other types of mass burning furnace used for coal and fibrous fuels, where a significant store of fuel exists on the grate or in the furnace, the boiler response to changes of grate speed and fuel bed depth is very slow by comparison to changes of air flow. In such cases, the main parameter affecting boiler output is the change of air flow with changes of fuel flow following relatively slowly. While the air flow may be controlled automatically, with attended boilers the fuel input, and hence stoker speed or fuel bed depth, may be controlled by hand without a significant influence on the boiler response to load changes. In such cases, cross limiting is therefore ineffective.

### 8.4.2.5  Flow measurement and air flow control

Differential pressure or pressure drop readings resulting from fluid flow through an orifice plate or other restriction vary as the square of the fluid velocity. It is therefore necessary to linearise flow signals before they can be added. This is carried out by the use of a square-root extractor. This process is particularly important where two or more fuels are to be fired simultaneously and the total air flow required has to be computed by the control system.

With combined fuel firing both fuels can be modulated together, but more often in such cases one fuel is a waste or by-product and will either be fired at a fixed rate or as it is produced, i.e. as available. Combined firing of a waste gas and oil is an example where the gas flow controller would be operated from the upstream gas pressure. A fall in gas pressure would indicate that insufficient fuel was available for the particular valve opening, and the valve would close to maintain the pressure at the set point. The oil would be modulated by the master steam pressure signal to maintain the heat input and hence the boiler load.

An interesting example of combined fuel firing is with drainage gas for coal mines, which is liable to wide fluctuation in both quantity and calorific

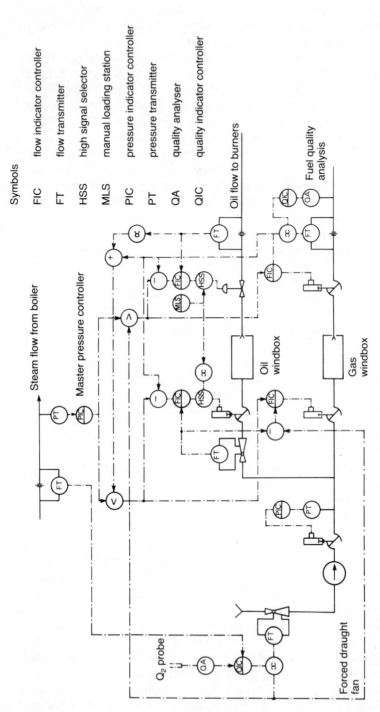

**8.7** Control diagram for oil- and gas-fired water-tube boiler

value. A gas calorimeter is used to measure the calorific value and to maintain the boiler pressure or temperature. Figure 8.7 illustrates such a system as used with oil and blast furnace gas.

A boiler may be capable of firing both oil and natural gas, but it is highly unlikely that they will be fired together since one or the other will be favoured at any one time due to economics and availability. Summation of the fuel flows need not therefore be included in such cases providing fuel changeover can be carried out manually.

To meet the requirements of the Clean Air Act, the gases emitted from the chimney must not exceed the degree of obscuration laid down. To ensure that these requirements are met it is therefore essential that the smoke density is measured. Optical smoke density instruments are arranged to give alarm in a situation when unacceptable conditions arise. Such instruments are essential with coal firing, desirable with oil firing, and of no use at all with gas firing.

## 8.5 Water level indication, measurement and control

### 8.5.1 Indication

This applies essentially to steam boilers where there is a visible water level. It embraces most steam boilers, the exception being those of the once-through or coil type, where there is no drum, in which case steam-outlet temperatures exceeding a preset value are taken to indicate a deficient water input. The principle of measurement and control is therefore similar to that described for hot-water boilers.

In most cases, however, the simple gauge glass (illustrated by Fig. 8.8) or, at high pressures, a double-plate glass variation (Fig. 8.9) on the steam/water drum or boiler shell is used as the indicator. Two must be provided[10] and arrangements made so that a breakage will not be a hazard to the operator. These arrangements usually take the form of balls which, in the event of glass failure, are forced over the steam and water apertures thus preventing the flow of these to the atmosphere. The instrument should also be easily read from the operating level, using periscopes or television if necessary, and be adequately illuminated. This time-tested device is used on the vast majority of boilers and is arranged to give a visible range of water level of ± 125 mm from normal water level.

There have, in recent years, been developed more sophisticated devices[11,12] which involve the use of electrodes, which are sensitive to water level. These may be of either the capacitance or the conductivity type. As well as providing a very clear visual display, the fact that they use electricity enables them to generate signals from which controls and alarms

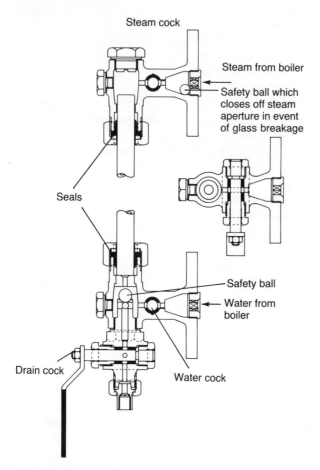

Steam cock

Steam from boiler

Safety ball which
closes off steam
aperture in event
of glass breakage

Seals

Safety ball

Water from
boiler

Drain cock

Water cock

**8.8**   Water gauge glass
(From Hopkinson Limited)

can be operated. Figure 8.10 illustrates a typical design. These devices are naturally more costly than gauge glasses but have many advantages – they are robust, self-illuminated, very easy to read, and are suitable for remote indication or recording. Moreover, they are ideal for use at very high pressures and temperatures.

Where the water gauge glasses are very distant from the operating level, remote level indicators using mercury-filled manometers or diaphragms are used to give a readily visible indication of boiler water level (Fig. 8.11). These usually give a visible range of ± 250 mm from normal water level.

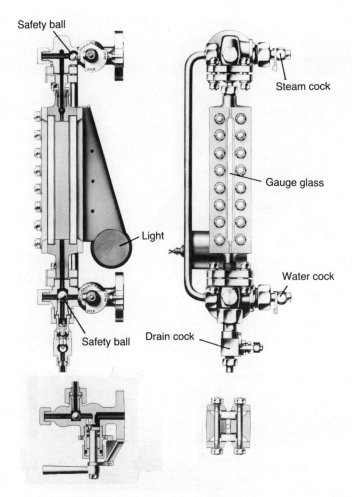

**8.9**   Double-plate glass water level indicator
(From Hopkinson Limited)

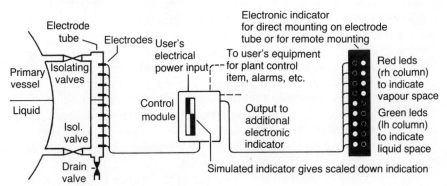

**8.10**   Electronic water level indicator
(From Hopkinson Limited)

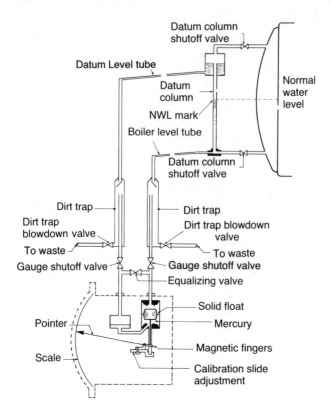

**8.11**   Remote water level indicator
         (From Babcock-Bristol Limited)

### 8.5.2   Measurement

The obvious way to sense water level is to use floats; these have been
successfully used for over fifty years to generate signals for shut-down in
the event of low water level and an on-off or modulating signal for
feedwater control (see Fig. 8.1). Such single-element controls are still very
widely used on industrial steam boilers. In small boilers, the floats may be
placed in the boiler shell, but in most cases they are located in small
external chambers. Two such chambers are mandatory, one initiating the
low level alarm and shut-down signal, the other initiating feedwater control
and, in some cases, the high level alarm. The top of each chamber is
connected to the steam space, the bottom to the water space and to drain,
the latter via a valve which is normally closed.

   The movement of the float is transmitted to the outside of the chamber
by means of a magnetic coupling. This operates switches to generate the

desired signals. For a modulating signal, switching causes the feed valve to open slowly on the water level falling, the movement of the valve being fed back to the switch as the valve opens to cancel the signal. This proceeds until the demand for feedwater is satisfied. Alternatively, the float movement can be transmitted by inductive coupling (Fig. 8.1).

It is an important requirement with water level controls that they be tested regularly, to which end they must be blown down to verify that they operate correctly and to remove sludge which could interfere with the free movement of the float. The operation may be performed by using a 'sequencing blowdown valve' which safeguards the correct procedure and eliminates operator intervention. This is illustrated in Fig. 8.1 as the automatic control check system.

These float controls have given years of good service but there is now a trend towards the sensing of water level by electrodes, which can be of either the conductivity or of the capacitance type. As explained in section 8.5.1, these can be incorporated in the display device but, at the time of writing, it is more usual for simple gauge glasses to be used, and the electrodes placed within the boiler shell (for small boilers) or in separate chambers as with floats. Whether floats or electrodes are used in chambers, it is important that the two chambers which must be used are entirely independent of one another, hydraulically and electrically.

### 8.5.3  Control

There are three basic forms of water level control used. These are single-element, two-element and three-element systems.

Single-element control adjusts the feedwater valve position in response to changes of boiler water level only (see Fig. 8.12(a)). When the load on a boiler is increased, there is a fall in pressure and an increase in firing rate, both of which result in a rise in water level in the boiler which is caused by the increased volume of steam below the water level. With single-element control such an action results in the feed valve closing, giving a decrease in water flow to the boiler when in fact there should be an increase to compensate for the increased steam flow. The level takes some time to fall to the desired position and will continue to fall below the required level before corrective action is taken. The reverse occurs on reduction of load. This system is quite satisfactory on firetube boilers with a large water capacity and with a reasonably steady load, but where fluctuations occur, and with a small water capacity, a very unstable water level can result.

This problem is overcome by the use of a two-element system (see Fig. 8.12(b)) which incorporates a signal from the steam flow at the boiler outlet and uses this as the primary signal for valve modulation. With an increase of steam flow the feedwater valve opens and increases the water flow proportionally. The drum level signal trims the valve position so as to

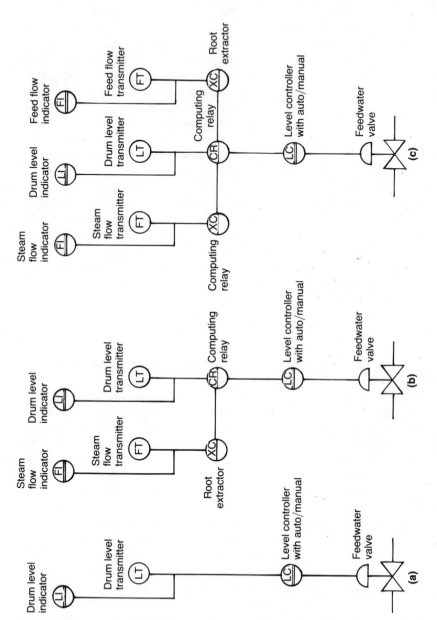

**8.12** Feedwater/drum level controls: (*a*) single-element control; (*b*) two-element control, (*c*) three-element control

regain the desired water level. The response of a two-element system is much more rapid than that of a single-element system since the change of steam flow is sensed immediately it occurs, giving an improved response to reasonable rates of load change.

For boilers subject to rapid load changes and for the larger industrial boilers, a three-element control is used which also incorporates a signal from the feedwater flow (see Fig. 8.12(c)). This system is very useful where feedwater pumps are common to a number of boilers, resulting in a variable feed-main pressure, and also where wide and rapid load changes occur. The feedwater flow gives a signal which is compared with the steam flow signal and compensates for changes of system characteristic which occur as the feedwater pressure changes with total flow following the feedwater pump characteristic.

The difference in cost between the various systems is influenced by whether or not steam and feedwater flow measurements are included as an instrument requirement. If they are, the same signal can be used for feedwater control and flow indication/recording.

## 8.6    Furnace pressure control

Firetube boilers and many water-tube boilers, particularly shop-assembled units fired by oil or gas, are operated with forced-draught fans only, the furnace pressure being always positive, often around 10 mbar, depending on the back pressure, positive or negative, of the exhaust system and heat-recovery equipment if fitted. Furnace pressure is allowed to vary with fuel and air input conditions; furnace pressure controllers are therefore rarely needed. This is known as pressurised furnace operation.

It is a different matter, however, when coal firing on stokers is used. A positive furnace pressure will cause products of partial combustion to percolate back through the coal supply into the boilerhouse, probably igniting the incoming fuel *en route*, which can be a serious event. All balanced-draught boilers therefore operate with a slightly negative furnace pressure, which should be kept constant at about 1 mbar sub-atmospheric. An induced-draught fan must therefore be used in addition to a forced-draught fan to create the necessary negative pressure. Excessive negative pressure will cause uncontrolled air to enter the furnace and interfere with good combustion. To avoid this, a furnace pressure controller is used.

The condition in the furnace is monitored and the resulting pressure signal passed to the furnace pressure controller. This is known as the 'measured value'. At the controller this measured value is compared with the preset desired value or 'set point'. Should a difference arise between the two values the 'error signal' passes to the positioner which regulates the movement of the actuator so as to effect the necessary adjustment, resulting in the damper of the fan opening or closing to reset the suction of the fan and hence the conditions within the furnace. Such a scheme is illustrated by Fig. 8.2. Alarms can be generated by the inclusion of monitoring devices set to close the contacts at specific levels.

The controller could consist of a diaphragm acting against a spring with atmospheric pressure on one side and furnace pressure on the other. Movement of the diaphragm causes a switch to operate a servomotor to move the induced draught damper towards the open or closed position, thus maintaining a near-constant furnace pressure. Firetube boilers using coking stokers (see Ch. 14) require induced draught only, furnace pressure being allowed to vary with demand for heat, by up to about 25 mbar. These do not require furnace pressure controllers.

## 8.7   Temperature measurement

The measurement of exit gas temperature in boiler plant is important in that, together with oxygen measurement, it is a factor in the calculation of flue gas heat loss (Ch. 11). A gradual rise in exit gas temperature for a given load on the boiler indicates boiler fouling, an increase of 17 °C signifying a heat loss of about one per cent. Typically, exit gas temperatures are in the order of 190–220 °C. Thermocouples, preferably bare wires supported in, but insulated from, the gas sampling tube, are used. They should be located in the mid-stream of the gas flow as near as possible to the actual boiler or heat-recovery equipment outlet. The read-out can be merely by indication. In modern instruments this is digital, but in addition a recorder should be used, as this shows trends with time and makes it easy to determine when the boiler should be taken off-load and cleaned.

In firetube boilers a thermocouple placed in the reversal chamber well above the furnace exit is useful with all forms of fuel firing, but especially with coal, when temperatures exceeding 900 °C can accelerate deposit formation on the tube inlets. The input to the boiler should be reduced if excessive temperatures be noted, and in turn this is indicative that the boiler needs to be cleaned.

Where superheaters are fitted a thermocouple in the steam outlet should be used. A decrease in steam temperature at a given load signifies fouling of the superheater (this is common with both coal and oil firing) and the need for boiler cleaning.

With hot-water boilers it is necessary to know both flow and return temperatures. If the latter should be below about 130 °C there is a danger of corrosion if fuel oil is being fired (Ch. 10). If indication alone is required, mercury-in-steel thermometers may be used, but in the modern context thermocouples giving a digital read-out and record are more desirable.

## 8.8   Boiler water quality

It is important that the quality of water within the boiler be kept within the limits recommended in BS2486 (see Ch. 6). This is generally carried out by sampling and testing the water on a regular basis: full instrumental measurement and control have not yet arrived. The measurement of the conductivity of the water is relatively simple and reliable, using electrodes, the water between them acting as one arm of a Wheatstone bridge. The conductivity of the water is, broadly, a measurement of the total dissolved solids content of the boiler water, which is an important property. This should be limited as indicated by the boiler vendor, or BS2486 where, in Table 2 referring to firetube boilers, the recommended limit for TDS is 3500 p.p.m. Table 3 of that standard refers to water-tube boilers and recommends TDS up to a maximum of 3000 p.p.m. according to boiler pressure.

The conductivity instrument can be arranged to trim the discharge from the continuous blowdown system, should this be fitted (which is not always the case), otherwise the display should guide the operator to adjust his manual blowdown regime as necessary.

Measurement of pH gives an indication of alkalinity. Normally, this should be about 9: higher figures indicating high alkalinity with the likelihood of priming; low figures warning that the water in the boiler could be corrosive.

## 8.9   Steam temperature control

The need to maintain a constant steam temperature at the superheater outlet of superheated steam boilers is dictated by the requirements of the steam user. Boilers very rarely give a constant temperature over their normal operating range (see Fig 2.11). Therefore, if a constant temperature is required over a given load range, it is necessary to include steam temperature control equipment (see section 2.5.1).

The control system is similar no matter whether direct or indirect desuperheaters are used. The controls can be of single-, two- or three-element types. If the desuperheater is placed at the outlet of a single-stage superheater, or if the boiler load is reasonably steady with the desuperheater between two stages of superheater tubes, a single-element control using the steam temperature at the superheater outlet as the only source of signal for operating the desuperheater controls will suffice.

With frequent or rapid load changes on a boiler having an interstage desuperheater (see Fig. 2.12) there is a delay between the operation of the desuperheater and the effect of this being sensed at the outlet of the secondary superheater due to the distance the steam has to travel between the desuperheater and the temperature sensor. To overcome this delay an additional signal (second element) is introduced from the final superheater inlet temperature.

For three-element control, a signal of boiler load is introduced which gives an anticipatory feed-forward signal to the controller which indicates the need for a change of position of the desuperheater controls, and this is useful for boilers subjected to very variable loads.

## 8.10   Combustion safety (burner management)

A burner management system incorporates interlocks to ensure that the correct sequence of events is carried out during start-up, operation and shut-down of the boiler. The burner management system is in addition to the automatic controls but is interlinked with the various control systems and may use signals from this in performing its function. There are a number of codes of practice and guidelines for safety interlocks and pre- and post-ignition purging for the various systems.[13,14] Burner management, as with control systems, can vary in complexity. A simple system is where the operator manually positions the various dampers and valves in the correct sequence during the start-up procedure. Interlocks are included to prevent one stage being carried out if the previous stage is incomplete. For most modern boilers with gas or oil burners, a fully automated system can be incorporated in which the start-up procedure is initiated by the press of one button.

The complete system will normally be incorporated in a burner management panel and include lamps and alarms which indicate that the various steps in the sequence have been satisfactorily carried out, or that a particular interlock has been the cause of a trip. The alarms indicate to the operator the item causing the trip, hence assisting rapid fault diagnosis.

One of the major features of a burner management system is the use of scanners that detect the presence of ignition and of the main flame of individual burners with oil, gas and pulverised fuel firing. These ensure that, in the event of a flame failure, the burner or burners concerned are shut-down. On a single burner installation this means a boiler shut-down until the fault has been rectified. Under no circumstances should a safety device be bypassed however urgent the demand for steam or heat may be; fatal accidents have occurred from such practice.

## 8.11   General comments

Flow recorders can be arranged to integrate and give total flows over a period of time as well as the instantaneous values shown on flow indicators. This assists with overall heat balances and operational costing. Pressure, temperature and flow transmitters are mounted local to the point from which the measurement is taken. Instruments and controllers are usually grouped into panels placed at the operating position, which may be at the front of the boiler.

In large industrial installations, central control rooms are often used where the master controls for the boilers, turbines and process plant are all grouped. Local panels are also incorporated to enable some or all of the functions to be controlled locally during start-up or emergency. With multiple boiler installations, separate panels local to each boiler can be supplied or, preferably, a common panel on which all the instruments and controls are grouped to enable one operator to supervise all the boilers. This can result in a reduction of the number of boiler-house personnel required and make it easier to gain an appreciation of the total plant state.

Control valves and dampers rarely give complete shut-off. Moreover, at low flow rates, poor mixing of fuel and air causes combustion problems. Thus, the ideal of infinite turndown is impossible to achieve. The turndown ratio $\left( \dfrac{\text{maximum continuous load}}{\text{minimum load on automatic control}} \right)$ is dictated by the ability of the various valves, dampers and speed regulators to maintain steady conditions for a given position. Generally, a control valve and damper will have a better turndown capability if it has a high pressure drop across it at full load. Rotating louvre dampers have a notoriously bad turndown. From $0°$ to about $40°$ of opening a large change of angle gives a small change of flow area, making accurate positioning difficult. This is particularly the case if all the blades rotate in the same direction. The performance can be improved by having pairs of contra-rotating blades with less than $90°$ rotation from fully open to fully closed. In practice, a turndown ratio on a single burner installation of 4:1 or 5:1 is achievable with oil or gas firing; with coal firing 3:1 is possible. This last ratio is adequate for most purposes for which firetube boilers are used. Water-tube boilers are rarely of single-burner type. Multiple burners enable higher turndown to be obtained by taking burners out of service at low load.

Control systems put forward for a particular application will vary in detail between system suppliers, depending upon the type of equipment they produce and upon the views of the design engineer concerned. The diagrams shown in this chapter can therefore only be indicative. The specialists should always be consulted and utilised to develop overall control schemes. It is essential, however, that an equipment purchaser appreciates what controls his system requires, in order to avoid the installation of unnecessarily complex schemes and high cost.

It is essential to appreciate that instrumentation and controls of all types require expert and regular servicing, otherwise the considerable benefits that their use can effect will be lost. It is therefore important that either a competent instruments and controls engineer is available within the purchaser's organisation or that maintenance facilities, provided by the supplier or his agents, are readily available at short notice. A regular service contract with such an organisation can be arranged, and is highly desirable.

## References

1. *Proposed Pressure Systems and Transportable Gas Containers.* Regulations and Approved Codes of Practice, HSC. HMSO.
2. BS 1646: *Symbolic Representation for Process Measurement Control Functions and Instrumentation.* British Standards Institution, London.
3. Fuller, J. A. *The History of Safety Valves Leading to the Development of British Standards for Safety Valves*, 1984. Inst.Mech.Eng. Conference on Safe Pressure Relief.
4. BS 6759: *Safety Valves, Part 1. Specification for Safety Valves for Steam and Hot Water.* British Standards Institution, London.
5. *Automatically Controlled Steam and Hot Water Boilers.* Guidance note PM5, Health and Safety Executive, London.
6. *Requirements for Automatically Controlled Steam and Hot Water Boilers.* Associated Offices Technical Committee, St Mary's Parsonage, Manchester M60 9AP.
7. BS 1042: *Fluid Flow in Closed Conduits. Part 1. Pressure Differential Devices.* British Standards Institution, London.
8. *Shell Flow Meter Engineering Handbook*, Sections 3 and 7.
9. Ormerod, W. G. *Automatic Trimming of Combustion Air to a Pulverised-Coal Fired Boiler Using the Flue Gas Carbon Monoxide Signal.* I. Chem. E., 1981 (Sept.) p. 174.
10. *Factories Act 1961*, Section 34.
11. Trade literature, Hopkinsons Ltd, Huddersfield HD1 2UR.
12. Trade literature, GESTRA (UK) Ltd, Hitchin SG5 1PH.
13. *Code of Practice for the Use of Gas in Low Temperature Plant.* British Gas publication IM/18.
14. National Fire Protection Association (NFPA), USA.

# CHAPTER 9

# Safety and environmental requirements

## 9.0 Safety requirements

In Chapter 1 the development of boilers was described, particularly that
taking place in the last century when low-pressure boilers gave way to
those operating at much higher pressures. This period coincided with many
catastrophic boiler explosions, with consequent deaths and injuries.
Coroners' inquests, then in ignorance of engineering facts, often ascribed
these disasters to 'Acts of God'.

In the middle years of the nineteenth century Manchester had the largest
concentration of steam boilers anywhere in the world, and it was natural
that efforts to stem the tide should begin there. The Manchester Steam
Users Association was formed with the object of regular inspection of
boilers and their ancillary equipment. Robert Longridge was appointed
Chief Inspector in 1855 and later argued that inspection should be
supplemented by insurance. The Steam Boiler Assurance Company was
formed in 1859, Longridge becoming Chief Engineer. The success of this
policy, after five years' operation was summarised by Longridge as
follows[1]: 'of 10 818 boilers insured by the company only 8 exploded,
whereas in the same period amongst uninsured boilers no less than 261
exploded with a loss of 498 lives, serious injuries to larger numbers, and
great destruction of property.' Insurance and inspection are today
mandatory by law. William Fairbairn was contemporary with Longridge
and it is largely due to his work that the Boiler Explosions Act was passed
in 1882. Fairbairn also carried out numerous experiments related to the
stresses in boilers, and some of his formulae formed the bases for design
calculations in modern standards. He was later knighted for his efforts.

## 9.1  Legislation

In the UK all pressure systems working above atmospheric pressure and containing steam or pressurised hot water are subject to regulation under The Factories Act 1961 (Ch. 3A). Certain sections are likely to be replaced by Regulations and Approved Codes of Practice under the Health and Safety at Work, etc., Act 1974[2], under which, among other factors, the user is required to have 'a written statement of the safe operating parameters of the system and a written scheme for the examination of the system by a competent person, and to have both these documents endorsed as satisfactory by a competent person before the system can be operated.' Although these regulations are only in draft form at the time of writing, readers, be they the boiler manufacturer, installer, or user would do well to be aware of their legal obligations by studying both Chapter 4 of the 1961 Factories Act and the proposed regulations which seem likely to replace it. The term 'competent person' is defined in the Proposed Regulations, but, for the present it may be taken to mean a suitably qualified person usually employed by an independent authority such as an engineering insurance company. Subsequent examinations of the system, which include boilers, must take place at intervals not normally exceeding fourteen months, but this period may on occasions, be varied according to the nature and use of the plant.

Apart from the explosive potential of a boiler as a pressure-containing vessel, there are various other aspects to safety such as: explosion or accidental fire on the fuel side of the boiler; possible dangers in obtaining external or internal access to the plant; the risk of burning due to human contact with hot parts; and the leakage of noxious gases into the boiler-house. Various standards and codes of practice have evolved to minimise the risk, and some of these are likely to be referred to in legislation in the UK and elsewhere.

## 9.2  Standards

In the UK the master pressure vessel standard is BS5500:1985 'Unfired fusion welded pressure vessels'. Boilers are pressure vessels, generally of complex construction, which are complicated further by being subject to intense heat in parts, which gives rise to stressing and corrosion which are not present in simple pressure vessels. Special constructional standards for boilers are therefore needed. The British Standards are the following.

BS855:1976:   'Specification for welded steel boilers for central heating and indirect hot water supply (rated output 44 kW to 3 MW).' Such boilers come within the scope of this book where maximum operating conditions are up to 4.5 bar and 132 °C for hot water and up to 2 bar, 132 °C and 1.5 MW for steam, the latter category being very rarely used.

BS2790:1986: 'Specification for shell boilers of welded construction.' This applies to all firetube boilers for duties beyond the limits of BS855.

BS1113:1985: 'Watertube steam generating plant (including superheaters, reheaters and steel tube economisers).' This standard applies to the plant described in its title, which includes superheaters and economisers for firetube boilers.

The last two standards are in the modern loose-leaf format, which allows automatic annual updating.

A boiler comprises not only the pressure vessel, but also ancillary equipment such as valves, gauges, pipes and flanges. It is also made from specified materials and is subject to various forms of testing during and after manufacture. All these items are governed by separate standards which are referred to and listed in the main standards, e.g. BS2790 contains a list of 36 other British Standards which are related to it. Other British Standards which are of importance in connection with boilers are: BS799:Part 4, 'Atomising burners for over 36 l hr$^{-1}$ and associated equipment for single and multi-burner installations'; Part 5, 'Oil storage tanks'; Part 6, 'Safety times and safety control and monitoring devices for atomising oil burners of the monobloc type'; and BS5885:1980: 'Specification for industrial gas burners of input rating 6 kW and above'. The six parts of BS5750 'Quality Systems' are likely to be related to boiler plant quality control in the future.

There are corresponding standards in most countries throughout the world: some accept British Standards; others DIN (West German); others use ASME (American Society of Mechanical Engineers); and many have their own. Continuous efforts are being made to harmonise standards; in Europe by the EEC using Special Directives, and world-wide by the ISO (International Standards Organisation). Good though the intention may be, progress is slow. In the UK the use of standards is not yet mandatory, but it could be useful in defence to show that a boiler and its auxiliaries have been made and installed in accordance with the appropriate standards and codes of practice. In many countries compliance with national standards is obligatory.

## 9.3 Safety in operation

Having built and installed a boiler which is inherently safe as a pressure vessel, it must be ensured that it is safe to operate as a steam or hot-water generator.

## 9.3.1 Water level

Modern industrial boilers fired with oil or gas operate with minimum manual attention; they are in effect fully automatic. Coal-fired boilers are also capable of a high degree of automatic operation but need may arise for occasional attention to the fires and to the fuel- and ash-handling systems. In all cases, plant operation can be monitored and controlled from a distance, any faults disclosed being rectified by the dispatch of a trained person to the boiler-house. Increasingly, boilers are now operated by energy management systems using computers rather than by people. With steam boilers operating in the UK, however, it is mandatory that certain manual operations be carried out once per day or once per shift, and others be carried out weekly.[3,4] These relate to the observation and blowing down of gauge glasses, checking the operation of water level controls by blowing these down (Fig. 8.1), checking the operation of override low-level cut-offs (which shut the boiler down if the water level falls below a preset level). The weekly test requires that the water level in the boiler shall be lowered to prove the operation of the controls whereas, with the daily or shift test it is sufficient to blow down the external chambers in which the level sensors are mounted. All tests must be carried out by a suitably trained person, who must ensure that all operating valves are returned to the operating position after use. Manufacturers of controls now supply equipment to monitor the operations and to ensure that they are carried out (Fig. 8.1). After the weekly test the operator must remain with the boiler for at least twenty minutes to ensure that its operation is correct.

These operations should be carefully logged. It is essential that boiler operators be informed of these and other requirements in references 3 and 4, which should be carefully studied by the reader.

## 9.3.2 Ignition

For many years ignition of gas and oil burners has been automated, using an electric spark to ignite the flame with or without the intermediate assistance of a gas pilot. Ignition is proved by a photovoltaic cell; once the main flame has been proved both spark and pilot are extinguished. Should ignition of either the main flame or pilot not be proved the controls must effectively shut off the fuel supply and require manual resetting.[3,4] The air supply remains for a short time to provide a 'post purge' of the gases in the boiler[5]. The resetting operation incorporates a 'pre-purge' which will enable the ignition process to be initiated without fear of an explosion occurring in the gas passages of the boiler. It is essential that the fuel shut-off valves are duplicated in series and are, if possible, proved shut. This is mandatory with gas[6], but more difficult to accomplish with oil. With gas oil cases have been known where leakage through a single shut-off valve (in former days; they are not permitted now) resulted in the accumulation of relatively low flash-point oil in a hot furnace, explosion occurring immediately the spark was initiated.

### 9.3.3 Water-side fouling

Water-side deposits can cause overheating of the pressure parts of a boiler and their eventual failure; therefore correct control over the composition of the water in the boiler is essential. This has been discussed in Chapter 6.

### 9.3.4 Hot surfaces

External hot surfaces are a hazard to persons who may accidentally come into contact with them. The nature of the surface is also important, at a given temperature the flow of heat from a dense material like steel to the human skin is much more rapid than from a less dense material such as thermal insulation; human tissue will be severely damaged if it reaches a temperature of 43 °C. Corresponding 'safe' surface temperatures for various materials have yet to be established, the present guideline being to protect or insulate all hot surfaces which can foreseeably cause damage to humans. This requires careful judgement not only in connection with the boiler structure but also steam and water pipes, ductwork, access doors, etc., in the environment of the boiler.

### 9.3.5 Access

Safety of access to boiler parts during operation and for internal inspection is important, the latter being considered in the constructional standards.[7,8,9] Access during operation to valves, gauges, controls and auxiliaries is not yet the subject of standards or Health and Safety Executive guidance notes. It is best to discuss proposals with local factory inspectors, but some guidance, not specific to boiler plant may be found in reference 10.

## 9.4 Boilers and the environment

### 9.4.1 Smoke

Most boilers were fired by coal until the early 1950s, many were still hand-fired on crude grates, smoking chimneys were usual in urban areas, and dense 'pea-soup' fogs occurred in due season. All these had been tolerated since the Industrial Revolution. In December 1952 a disastrous fog settled over London, resulting in about 4000 deaths over four days.[11] Public outcry prompted an energetic MP, Gerald Nabarro (later knighted), to propose a private member's bill to Parliament to control chimney emissions, but this was replaced by a Government Bill acting on two reports from a committee set up under Lord Beaver. The Clean Air Act was passed in 1956; it prohibited the emission of 'black smoke' (Ringelmann No. 4) for a period of more than two minutes in thirty, and of 'dark smoke' (Ringelmann No. 2) for more than four minutes, except for a certain relaxation for cleaning fires and soot blowing. The emission of particulate matter was also controlled.

The effect of the Act was almost immediate, 'pea soupers' are a thing of the past and one can see now across London or Manchester to the hills beyond. The 1956 Act was added to by another Act in 1968 which extended controls and brought emissions from liquid fuels under law. Reference 11 is a clear and concise digest of the Clean Air Acts.

In the early 1950s much of the problem arose from low-level emissions from domestic fires burning coal, but as much again originated from industry. Obviously, the nuisance can be mitigated by raising the level from which smoke is emitted. Nonhebel and Hawkins[12] studied the dispersal of smoke from industrial chimneys, and Bosanquet[13] the plume rise over the chimney exit.

From these and other works, a Memorandum on Chimney Heights[14] has been prepared by the Department of the Environment in connection with the 1956 Act. This provides a means for calculating the chimney height, for various fuels, and for various environmental conditions. This work is based on the emission of sulphur dioxide rather than of smoke, this gas being the main irritant in the products of combustion.

### 9.4.2 Oxides of sulphur and nitrogen

The cleansing of air is not yet complete. Having virtually eliminated smoke there is still the problem of sulphur dioxide ($SO_2$) and, more recently, that of oxides of nitrogen ($NO_x$). This latter is thought to cause the smogs of Los Angeles and Tokyo, and also to contribute to the European and Scandinavian troubles with acid rain, about which there is much controversy. Martin and Barber[15] have described the monitoring of $SO_2$ and $NO_x$ in rural England. The EEC has a directive in draft which will set onerous limits on the emission of these gases from plants with installed capacities exceeding 50 MW; this embraces many installations of multiple firetube and most water-tube boilers. This looks like setting boiler designers, and users, problems for many years to come. Sulphur dioxide may be controlled by scrubbing the exhaust gases, which raises many problems, or by retention of sulphur in fluidised beds (see Ch. 14). Nitrogen oxides are more difficult to deal with, as some of these are produced from nitrogen in the fuel and some from nitrogen in the air. Extensive research is needed before solutions can be obtained. This is progressing in the context of power-station boilers.[16] Staged combustion, low excess air, flue gas treatment and catalysis are all being considered. The problem of $NO_x$ control in the smaller industrial boiler does not yet appear to have been tackled.

### 9.4.3  Grit and dust

Grit and dust or fly ash will obviously be emitted from plant burning solid fuel, the smaller the fuel particles and the greater the ash content, the greater is the emission. Heavy fuel oil, even though the ash content is low (<0.1%), can emit particles of carbon ('stack solids') if combustion conditions are unfavourable (see Ch. 4). Grit consists of particles greater than 75 $\mu$m diameter. Dust consists of particles between 75 and 1 $\mu$m diameter. There is a further category, 'fume', the legislative position of which has not yet been resolved. This refers to particles of less than 1 $\mu$m diameter.

The present UK law requires that the emission levels be limited according to the size of the plant and the fuel used, for example, with coal firing from a plant of 1.46 MW input the permitted emission level is 1.95 kg hr$^{-1}$, representing about 1% of the fuel fired. For a large plant of 87.9 MW input the permitted emission is 60 kg hr$^{-1}$ or 0.55% of the fuel fired. With oil firing the restriction is more severe: 0.5 kg hr$^{-1}$ for the small 1.46 MW plant (e.g. about 0.4% of the fuel fired); and 35 kg hr$^{-1}$ for the larger 87.9 MW plant or 0.2% of the fuel fired.[3] For more detailed information about permitted emission levels for solid particulates the reader should consult, in the first instance, reference 3 and then the regulations connected with the Clean Air Act.[17,18] It is emphasised that the present requirements are likely to be tightened in order to comply with EEC regulations. With coal firing, grit arrestors must be used to comply with the present regulations.

#### 9.4.3.1  *Dust collection equipment*

The quantity of grit and dust produced within a boiler depends upon many factors, the most influential being the combustion process used; others are the fuel density, fineness, ash content and moisture content. The quantity discharged from a boiler is also influenced by the gas flow pattern over the boiler heated surfaces. This may be arranged to encourage fall-out of dust into boiler hoppers under the action of gravity and changes of the direction of the gas flow.

The quantity of dust carried in the gases is referred to as the 'dust burden' or 'dust loading' and is expressed either as a total flow rate in $kg\ hr^{-1}$ or as milligrams per cubic metre of gas at specified conditions of pressure and temperature.

The particle size can vary from almost zero to well over 100 $\mu m$. The size distribution will depend upon a number of factors, among which are the sizing of the fuel burned and the combustion process used. The particle size distribution is expressed as the percentage of the total dust content passing a series of micron sizes or as the percentage of the total dust content in a series of size ranges (see Table 9.1).

**9.1** Methods of expressing dust loadings

| Particle Micron Size ($\mu$m) | Cumulative Percentage by weight through micron size |
|---|---|
| 150 | 85.0 |
| 75 | 68.0 |
| 50 | 50.5 |
| 40 | 41.0 |
| 30 | 32.0 |
| 20 | 24.0 |
| 10 | 15.0 |
| 5 | 9.0 |

| Range of Particle Size (μm) | Percentage in range by mass (= percentage through high size − percentage through low size) |
|---|---|
| > 150 | 100 − 85 = 15.0 |
| 75–150 | 85 − 68 = 17.0 |
| 50–75 | 68 − 50.5 = 17.5 |
| 40–50 | 50.5 − 41 = 9.5 |
| 30–40 | 41 − 32 = 9.0 |
| 20–30 | 32 − 24 = 8.0 |
| 10–20 | 24 − 15 = 9.0 |
| 5–10 | 15 − 9 = 6.0 |
| < 5 | 9.0 |
| | 100.0 |

As already explained there are limitations imposed by the Clean Air Act in the UK and various other national authorities upon the amount of grit and dust that can be discharged to the atmosphere from equipment burning fuels. As an example, a typical United Kingdom requirement is not more than 115 mg N $m^{-3}$ at 12% $CO_2$. Specifying the $CO_2$ content of the flue gases when expressing the dust loading in this manner prevents the specified concentration being achieved by the introduction of dilution air prior to the test point, thus increasing the gas volume and hence reducing the concentration per unit volume while still discharging the same total quantity of dust to the atmosphere.

The amount of carryover of solids for a given fuel is at a minimum with travelling grate stokers, increasing with the use of spreader stokers, up to maximum with pulverised fuel and fluidised beds. Similarly, the amount of fine material, as a percentage of the total, increases from travelling grates through to pulverised fuel firing and fluidised beds.

There are four major types of equipment used for removal of dust from flue gases, each having varying collection efficiency capabilities as follows:

| | |
|---|---|
| cyclones | efficiency range 85–95% |
| gas scrubbers | 95–98% |
| fabric or bag filters | 99% |
| electrostatic precipitators | 99% and above |

**9.2**  Typical calculation of overall efficiency for a cyclone collector

| Size ($\mu$m) | Percentage of total dust in range (stoker firing) | Efficiency of collector in range (%) | Percentage of total dust collected |
|---|---|---|---|
| | (1) | (2) | $((1) \times (2))/100$ |
| > 150 | 15.0 | 99.0 | 14.85 |
| 75–150 | 17.0 | 98.0 | 16.66 |
| 50–75 | 17.5 | 96.0 | 16.80 |
| 40–50 | 9.5 | 94.0 | 8.93 |
| 30–40 | 9.0 | 92.0 | 8.28 |
| 20–30 | 8.0 | 88.0 | 7.04 |
| 10–20 | 9.0 | 80.0 | 7.20 |
| 5–10 | 6.0 | 45.0 | 2.70 |
| < 5 | 9.0 | 1.0 | 0.09 |
| | 100 | Overall collector efficiency | 82.55 |

| Size ($\mu$m) | Percentage of total dust in range (pulverised fuel) | Efficiency of collector in range (%) | Percentage of total dust collected |
|---|---|---|---|
| | (1) | (2) | $((1) \times (2))/100$ |
| >150 | 1.0 | 99.0 | 0.99 |
| 75–150 | 5.0 | 98.0 | 4.90 |
| 50–75 | 10.0 | 96.0 | 9.60 |
| 40–50 | 8.0 | 94.0 | 7.52 |
| 30–40 | 13.0 | 92.0 | 11.96 |
| 20–30 | 13.0 | 88.0 | 11.44 |
| 10–20 | 20.0 | 80.0 | 16.00 |
| 5–10 | 15.0 | 45.0 | 6.75 |
| < 5 | 15.0 | 1.0 | 0.15 |
| | 100 | Overall collector efficiency | 69.31 |

Because the collection efficiency of a cyclone varies widely with particle size, it is essential if a statement of the overall efficiency of a collector is to be meaningful, that the inlet dust size distribution and density be specified to the supplier. Table 9.2 indicates how the overall collection efficiency is determined from the grade efficiency characteristics of the collector and the effect on this of the size distribution.

### 9.4.3.2  Cyclone

This is the simplest form of dust collector, but the efficiency of collection is low with small particle sizes (see Table 9.2, column (2)). The cyclone relies upon the centrifugal force imparted to the gases by rotation within the cyclone to separate out the dust, as illustrated by Fig. 9.1. High velocities therefore result in higher collection efficiencies than do low velocities but

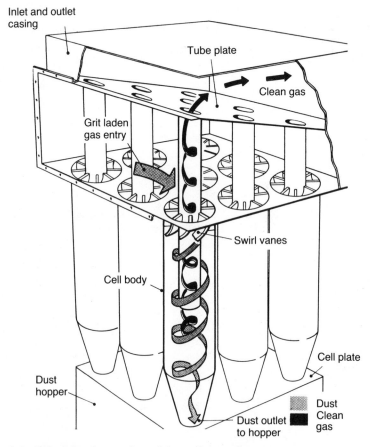

**9.1**  Principle of operation of the cyclone collector
(From Davidson and Company Limited)

they also give a higher gas-side pressure drop and hence higher induced-draught fan power consumption.

A wide range of variations of the basic cyclone exist and are still in the process of development. Most commercial collectors consist of multiple small cyclones of a standard configuration arranged in parallel on the gas side. For a given cyclone design therefore the number of cyclones is proportional to the gas flow.

The gas temperature at which a cyclone collector can operate is limited only by the material of construction, usually mild steel. The cyclone structure should be insulated to avoid condensation of the acid and moisture in the flue gas which can lead to clogging at low temperatures.

### 9.4.3.3   Gas scrubbers

These are devices in which the dust-laden gas stream is brought into contact with a water spray. There are medium- and high-energy types, the efficiency of mixing of the water and gas being a measure of the energy used and also of the collection efficiency of the scrubber. Depending upon the gas inlet temperature, gas scrubbers increase the volume of gas discharged to the atmosphere due to the volume of water evaporated, this can result in a visible plume at the chimney discharge even though there is a minimal solids content. This is often mistaken for smoke and hence dust emission.

A disadvantage of gas scrubbers is that there is a waste water effluent to be disposed of which contains all the collected solids. Separation of the solids from the water will have to be carried out to enable them to be transported for disposal; the cost of this equipment and the space required to accommodate it must be taken into account in the overall evaluation.

Gas scrubbers tend to be used with low-sulphur fuels, where the electrostatic precipitator is less effective, or where very hot gases are to be discharged to atmosphere without heat recovery. The water cools the gases in addition to cleaning them, reducing the temperature to a level acceptable to the induced-draught fans and chimneys. An example of such an application is a refuse incinerator without heat recovery. It should be appreciated that the plume rise will be lower with gases discharged at low temperature. This can be corrected by using a conical cowl at the chimney exit which will increase the induced-draught fan power so as to accelerate the gases.

### 9.4.3.4   Fabric or bag filters

These consist of a series of woven fabric bags arranged in parallel on the gas side, through which the gases pass leaving the dust retained on the bag material. The dust is removed by periodic vibration or shaking of the bags.

Due to the non-availability of suitable economic bag materials, the maximum operating gas temperature is limited to about 260 °C. Equally important is the lower temperature limit which must be selected to ensure that the gases do not operate at or below their dewpoint. This could result in damp or wet dust entering and blocking the bags and the bags being subjected to the effect of corrosive constituents.

The gas pressure drop across a bag filter tends to be high, and increases as the deposited load on the bags increases.

### 9.4.3.5 Electrostatic precipitators

These operate by charging the dust particles as the dust-laden gas passes across electrically charged wires. The dust is then attracted to, and collected on, oppositely charged plates which are periodically mechanically rapped or vibrated to remove the dust adhering to them. To achieve optimum collecting efficiency the velocity of the gases through the precipitator must be uniform across the flow path, and this requires careful design of the inlet and outlet ducting.

The collecting efficiency is not particularly affected by particle size. The gas-side pressure loss is low by comparison with other types of dust collector, thus compensating for the electricity consumption necessary to charge the dust particles. Care has to be taken to ensure that the precipitator does not operate below the acid dew point of the flue gases which would subject the materials of construction to corrosion.

Precipitators tend to be large and expensive when compared with most other collectors but they can achieve high collection efficiencies with fine materials.

With low-sulphur fuels producing high-resistivity grits, precipitators have to be very large to achieve the desired collection efficiency unless the sulphur content of the gases is artificially increased.

### 9.4.3.6 Combined collectors

The aforementioned types of collector can be used in combination to achieve the desired emission levels, e.g. precipitators may be preceded by mechanical cyclones to remove the larger dust particles. There have been occasions when a high-efficiency bag filter has been used on the gas side in parallel with a mechanical cyclone, where the latter cannot alone achieve the design emission level and a bag of adequate size is too bulky and expensive. The proportion of gas flowing through each is selected to give the desired combined efficiency. Such a combination can give a lower capital cost scheme than a bag filter only. This procedure can be of use as a retrofit, where the design emission has not been achieved by the originally installed cyclone, or where the fuel conditions have changed, thus necessitating a higher collecting efficiency to achieve the desired emission level.

## 9.4.4  Smutting

This used to be a common occurrence when heavy fuel oil was used. Heavy fuel oil has a high sulphur content. This oxidises to $SO_2$ and also, to a smaller extent, to $SO_3$, which combines with the water vapour resulting from the combustion of hydrogen in the fuel to form sulphuric acid (see also sections 7.3 and 10.4.2). Sulphuric acid will start to condense at relatively high temperatures, around 135 °C, and forms a sticky layer on any surface below this temperature. This will attract and retain particulate matter which will in turn tend to absorb the acid, thus forming a weak skin on the surface on which it settles. This skin will easily break up when disturbed by the pulse from an igniting burner or gust of wind, and then be carried out of the plant by the products of combustion. The flakes will settle locally and, because they contain sulphuric acid, will corrode and mark anything on which they alight.

The heated surfaces of the industrial boiler are mostly above about 135 °C (the 'acid dewpoint'), so that the phenomenon does not occur in the boiler itself but in the cooler parts which follow it, i.e. in the air preheaters, in the ducting and in the chimneys (see also sections 7.3 and 7.4). Smutting can be a serious nuisance. The main form of control lies in ensuring that no surfaces in contact with the products of combustion fall below the acid dewpoint except during start-up conditions. Chimneys, particularly the top section, which is subject to cooling by direct radiation loss on clear nights[19], internal downflow of cold air due to wind, and conduction loss through the main structure, are very likely to be sources of smuts. The top should be tapered[20] and insulated, and the main barrel insulated using a non-porous lining such as steel backed up with rock wool or fibreglass. It is important that acid gases should not percolate into and migrate through the insulation to the outer skin, where it could condense and eventually damage the main structure. Ravenscroft and Page[21] and Gills and Lees[22] have studied the heat-transfer conditions in chimneys and have provided information whereby the thickness and properties of the insulation required can be calculated. For general guidance on chimney design the reader is referred to *The Brightside Chimney Design Manual*.[23] Most proprietary chimney suppliers are now aware of the problems and how these may be controlled.

An alternative or additional approach to the control of smutting is to neutralise and dry up the condensed acid. This may be done by injecting magnesium hydroxide into the products of combustion upstream of the cool surfaces.[24,25] This is an effective method of smut control where the volume and temperature of the products of combustion are below the design capacity of the chimney, or where an old chimney of excessive diameter is used. Such cases can arise during periods of reduced output from a factory.

Usher and Shephard[26] have recently given an overview of proposed atomospheric pollution regulations for the UK, EEC and USA which should be studied.

### 9.4.5 Noise

Some auxiliary equipment of boiler plant can generate excessive noise; fans are particularly liable to do so. In larger installations it may be possible to locate this kind of equipment so that the noise problem is localised away from occupied areas; with the smaller 'package' boilers the noisy equipment is located on the boiler. Means whereby noise can be predicted and controlled to acceptable levels are well established[27,28], e.g. by using fan-inlet muffles and by insulating the casings, the former often being sufficient to meet the desired noise level.

Combustion noise is not usually a problem but it can arise through unstable combustion (see Ch. 4). Cases have arisen in firetube boilers where flame instability has resonated with the dimensions of the cylindrical furnace, resulting in an organ note of around 250 Hz being emitted from the chimney at a weighted sound pressure level of about 60 dBa. Normally, this would not be a nuisance but it can prove intolerable in a residential area at night. The cure is to trace and correct the combustion problem which is the driving force. Another phenomenon which sometimes arises from combustion is infrasound at about 4 Hz. This cannot be heard, but may resonate with the chest cavity of the human body, thus giving rise to feelings of nausea. The clue to such an occurrence may be found in the visible vibration of glass panels in the locality. The cure, as before, is to eliminate the instability.

Noise in boiler plant can also arise from coal- and ash-handling equipment, particularly the lean-phase pneumatic system, where coal is conveyed at high velocity in an air stream through steel pipes. Correct siting and anchoring of the equipment, the avoidance of sharp bends, and insulation seem to be the only means of control. Dense-phase systems, in which coal is conveyed as plugs in the pipework, are considerably less noisy than lean-phase ones.

## References

1. National Vulcan Engineering Insurance Group Ltd. *Vigilance* (Vol. 4), No. 5, Manchester. (1984) p. 12.
2. *Proposed Pressure Systems and Transportable Gas Containers Regulations and Approved Codes of Practice*. HMSO, London, 1984.
3. *Requirements for Automatically Controlled Steam and Hot Water Boilers*. Associated Offices Technical Committee, Manchester.
4. *Automatically Controlled Steam and Hot Water Boilers*. Guidance note PM5, Health and Safety Executive, London.

5. BS 799:1979 *Atomizing Burners for over 36 l h⁻¹ and Associated Equipment for Single and Multi-burner Installations, Part 6, Safety Times and Safety Control and Monitoring Devices for Atomizing Burners of the Monobloc Type.* British Standards Institution, London.

6. BS 5886:1980 *Specification for Industrial Gas Burners, of Input Rating 60 kw and Above.* British Standards Institution, London.

7. BS 855:1976 *Specification for Welded Steel Boilers for Central Heating and Indirect Hot Water Supply (Rated Output 44 kw to 3 MW).* British Standards Institution, London.

8. BS 2790:1986 *Specification for Shell Boilers of Welded Construction.* British Standards Institution, London.

9. BS 1113:1985 *Design and Manufacture of Water-tube Steam Generating Plant (Including Superheaters, Reheaters, and Steel Tube Economizers)* British Standards Institution, London.

10. 'Factory stairways, ladders, and handrails', *Handbook No. 7.* Engineering Equipment Users Association, London.

11. Short, W. and Harris, P. S. *The Clean Air Acts.* Bulletin of the Energy Users Research Association (associated with NIFES), London. 1984 (Apr.) **46**

12. Nonhebel, G. and Hawkins, J. E. 'Chimneys and the disposal of smoke'. *J. Inst. Fuel* 1955 (Nov.) **28**: p. 530.

13. Bosanquet, C. H. 'The rise of a hot waste gas plume', *J. Inst. Fuel* 1957 (June) **30**: p. 322.

14. 'Memorandum on chimney heights', *Clean Air Act 1956* (3rd edn). HMSO, London.

15. Martin, A. and Barber, F. R. 'Acid gases and acid rain monitored for over 5 years in rural East-Central England', *Atmospheric Environment,* Pergamon Press, London, 1984. **18.9**: p. 1715.

16. 'Nitrogen oxides from coal combustion', *Report No. ICTIS/TR11.* I.E.A. Coal Research, London.

17. *The Clean Air Act (Emission of Grit and Dust) Regulations 1971.* HMSO, London. S.I. 1971, **162**.

18. *Draft Regulation – Emission of Grit and Dust from Furnaces 1977.* Department of the Environment Ref. NPCA/366/1. 1977, London.

19. Pickles, J. H. 'Radiation heat loss from chimney tops', *J. Inst. E.* 1980 (Sept.) **53**: p. 416.

20. Anon. *Chimneys For Industrial Oil Fired Plant.* Shell UK Ltd, London. **24**.

21. Ravenscroft, R. P. and Page, H. A. 'Field investigation and design of insulated chimneys', *J. Inst. Fuel* 1966 (Jan.) **39**: p. 22.

22. Gills, B. G. and Lees, B. 'Design and operation of an insulated steel chimney', *J. Inst. Fuel* 1966 (Jan.) **39**: p. 29.

23. *Brightside Chimney Design Manual* (2nd edn.) Technitrade Journals Ltd, London.

24. *The Effective Way to Solve Acid Smut Problems.* Trade literature, Steetley Refractories Ltd, Hartlepool.

25. Conolly, R. and Kelsell, P. H. 'A direct assessment in 60 MW (e) oil fired boilers of the limitation of good practices in combatting acid deposition and associated smut emission and the consequent use of a neutralising additive'. *J. Inst. E.* 1982 (Mar.) **55**: p. 422.47.

26. Usher, S. M. and Shephard, F. E. 'Environmental Digest No. 12', *Gas Engineering and Management* 1984 (July/Aug.) p. 282. Inst. Gas. Eng., London.

27. Bolton, A. N. *'Predicting Fan Sound Pressure Levels'.* Conference Section, NEL, East Kilbride, 1976 (Jan.).

28. *Noise Reduction.* McGraw-Hill pp. 434–65, 541–70.

Chapter 10

# Fire-side deposits and corrosion

## 10.0   Occurrence

All commercial fuels, with the exception of most sources of natural gas, contain substances which can cause deposition and corrosion on the heated surfaces of boiler plant. Deposits can result in frequent shut-down for their removal to maintain the boilers in an efficient and operable state. Corrosion will result in the accumulation of products of corrosion and will make shut-down for cleaning necessary as well as forcing expensive repair work. The various aspects of this subject are very widely documented.[1]

It must not be thought that all fuels and all boilers will cause and suffer from deposits and corrosion. With careful design and selection the occurrence of these is exceptional rather than usual. Particular fuels, boilers, and operating conditions can all combine to cause or to avoid the problems. It is the purpose of this chapter to explore the mechanisms involved in deposition and corrosion and then to suggest methods by which they can be controlled.

## 10.1   Deposits, coal firing

### 10.1.1   Occurrence

Deposits occur in those parts of the boiler where the metal temperature is high, particularly on superheaters and their supports and at the entrance to convection heating surfaces following the furnace. They may occur in both water-tube and firetube types of boiler, to such an extent that boilers need to be shut down in a matter of weeks so that the deposits can be removed, otherwise the pressure loss caused by their obstruction of gas passages can rise beyond the capacity of the boiler fans. When this happens the boilers

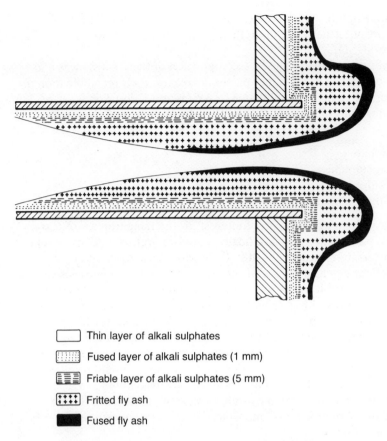

☐ Thin layer of alkali sulphates

▦ Fused layer of alkali sulphates (1 mm)

☰ Friable layer of alkali sulphates (5 mm)

▦ Fritted fly ash

■ Fused fly ash

**10.1** Section through a deposit on a tube inlet in a firetube boiler
(After W. M. Crane, BCURA Report No 254, 1962)

become virtually inoperable. Figure 10.1 illustrates a section through a
deposit at the inlet to a convection tube of a firetube boiler. It will be seen
that the reduction in cross-sectional area, and hence the increased
resistance to gas flow, is very considerable. Five distinct layers can be
identified, the second of which (counting from the metal surface) has
fused, thus capturing particles of fly ash which happen to alight on it. The
mechanism of deposition therefore involves a bonding process. It is the
sticky inner layers which cause the problem.

The main ash substances of coal are silica and alumina, which are not
easily fusible, but there are other substances present, e.g. iron, calcium,
sodium and potassium. It is the last two which play a dominant part in
causing deposits to adhere to heated boiler metal. Sodium and potassium
are members of the alkali group of metals; they may be present in coal as
chlorides, but this is not always the case. Another substance in the coal
which may be organically or inorganically combined with the coal or ash

substances is sulphur, which oxidises mainly to sulphur dioxide ($SO_2$), a small proportion, usually not more than 5%,[2] going further to the more active trioxide, $SO_3$.

The combustion process causes decomposition of these substances and, as temperature through the boiler passes decreases, their recombination in different forms, namely the sulphates ($NA_2SO_4$ and $K_2SO_4$) or pyrosulphates ($Na_2S_2O_7$ and $K_2S_2O_7$). The melting points of these salts are:

| | |
|---|---|
| Sodium sulphate | 880 °C |
| Sodium pyrosulphate | 401 °C |
| Potassium sulphate | 1069 °C |
| Potassium pyrosulphate | 300 °C |

It is the pyrosulphates which are of interest, as the temperatures of certain metal parts, particularly the superheaters, in both firetube and water-tube boilers can certainly exceed 300–400 °C, in which case there will be a molten or sticky layer on the metal surface to which fly-ash particles can adhere. As the deposit builds up, so its surface temperature will increase, and other substances of higher melting points can become liquid, as well as the initial low melting-point salts. As the deposits thicken, further reactions take place between silica, iron, sodium and potassium to give a molten slag which can be very dense. Where there are options, therefore, coals with ash substances low in sodium and potassium should be chosen, but it is an unfortunate fact that questions of fuel availability, combustibility, and price, often overrule this. Table 10.1 shows the ash analyses of two different coals, one which gave heavy deposition, the other which did not.

The difference in total alkali content is striking, and seems to be the main cause of the difference between the behaviour of the two coals. Some idea of the fouling propensity of the coal can be gained from calculating a fouling factor given by the expression:

$$\text{fouling factor}^3 = \frac{(Fe_2O_3 + CaO + MgO + Na_2O + K_2O)\, Na_2O}{SiO_2 + Al_2O_3 + TiO_2} \qquad [10.1]$$

Where the components are expressed as percentages of the ash. Referring to Table 10.1, the fouling factor for coal A is 0.7 compared with 0.1 for coal B.

Chlorine content has often been used as a quick indication of fouling, based on the assumption that the alkali metals exist in the unburned coal as chlorides. This is only partly true[4]; a coal with a chlorine content of 0.9% or above will always be fouling, one with a chlorine content of 0.3% may or may not be fouling. As has been previously stated, the alkali metal concentration which is the cause of the problem, may not always relate to the chlorine content.

**10.1** Ash analyses of two British coals

| Substances | Coal A, deposit forming | Coal B, deposit free |
|---|---|---|
| S in coal (%) | 1.53 | 1.73 |
| Ash in coal (%) | 7.5 | 7.4 |
| $SiO_2$ in ash (%) | 46.3 | 40.6 |
| $Al_2O_3$ in ash (%) | 26.8 | 33.5 |
| $Fe_2O_3$ in ash (%) | 15.4 | 10.8 |
| MgO in ash (%) | 1.3 | 3.0 |
| CaO in ash (%) | 2.7 | 6.7 |
| $TiO_2$ in ash (%) | 1.0 | 0.8 |
| $Na_2O$ in ash (%) | 2.2 } total alkali | 0.3 } total alkali |
| $K_2O$ in ash (%) | 3.1 }     5.3 | 0.8 }     1.1 |
| $SO_3$ in ash (%) | 0.7 | 4.0 |

## 10.1.2   Control of deposits from coal firing

There are several ways in which deposits and their effects can be controlled. These are given below.

### 10.1.2.1   Metal temperature control

Where possible avoid high metal temperatures. This cannot always be done, e.g. the superheated steam temperature will control the metal temperature of a superheater (section 5.9), and saturation temperature will control the metal temperature of evaporator surfaces. In firetube boilers, saturation temperatures are rarely such as to cause problems, but metal parts, especially the tube plate at the entry of the first pass of tubes, can be rather thick and this will cause high hot-face temperatures to occur. The projection of the tube ends into the gas space should be minimised and their internal edges well rounded. Tube plates should be as thin as the design will allow, and entry gas temperatures reduced as much as possible: they certainly should not be higher than 900 °C for coal, 650 °C for municipal refuse.

### 10.1.2.2   Correct tube spacing

In water-tube boilers and in all superheaters, avoid close-pitched tubes. This will reduce local gas velocities and hence mass heat transfer and, if the spacing is wide enough, prevent bridging. Wide pitching means that the heating surface required for given duty will be increased because the heat-transfer coefficient is reduced (Ch. 5). This means increased costs, but when this is offset against the cost of frequent outage and cleaning it can be well worth while.

### 10.1.2.3 Use of sootblowers

Use sootblowers strategically placed to remove the deposits. These are not always effective with strongly bonded deposits but can remove the more friable deposits once the initial layer has formed and before these have built up to an extent where the outer layers have fused. Sonic sootblowers are a recent development and in some cases are said to be very effective on account of their wide field of influence.

### 10.1.2.4 Fuel blending

It may be possible to reduce the fouling effects of a particular coal by blending it with one containing less deposit-forming substances such as sodium and potassium.

### 10.1.2.5 Chemical additives

Bearing in mind that the initiation of a deposit is caused by chemical reactions which form low melting-point salts, these processess or their results can be modified by chemical means using neutralising substances. Magnesium oxide, $MgO$, will react with the alkali pyrosulphates to form magnesium sulphate which does not melt, but decomposes at 1124 °C. The molten bonding substance is therefore eliminated. Furthermore, magnesium oxide has a very high melting point and will raise the melting point of any eutectics with which it is mixed. Magnesium oxide is easily available, generally being supplied in the form of magnesium hydroxide $Mg(OH)_2$, and it should be injected into the boiler as an airborne suspension, roughly in proportion to the total alkali content of the coal. In practice, very little is used, about 200 g $t^{-1}$ of coal being typical. The effect of magnesium oxide is to eliminate deposition entirely or to reduce it to a very friable substance easily removed by sootblowing or brushing.

Parts from which deposits have been removed should be washed down with water to bare metal. This removes remnants of the alkali sulphates which are water soluble. Washing, however, is not always feasible, as there may not be adequate drainage in the boiler for the removal of water if water washing was not anticipated at the design stage. Care must be taken when disposing of the wash water effluent as this may require treating before it can be put into a drainage system.

Fouling is detected by a rise in pressure loss (draught loss) across the gas passes of the boiler. It is normal for this to increase slowly with time, but if the rate of increase in pressure loss starts to rise, the boiler should be shut down and cleaned, otherwise it will shut itself down! For this reason, differential pressure gauges would be advantageous if fitted to the gas side of coal fired boilers.

## 10.2  Deposits, oil firing

### 10.2.1  Occurrence

Heavy residual fuel oil contains very little ash compared with coal (about 0.1% compared with 5% and more), but the nature of this ash is very different from that of coal ash, the constituents generally having lower melting points. Sodium and sulphur are present, the latter in greater degree than in coal ash, so that the mechanism of initial bonding, i.e. the formation of sodium pyrosulphate, will be the same as that for coal firing and the means of control will be similar. This is not all, however, as, owing to the low melting points of the ash constituents, direct impaction of some of these substances in a molten or semi-molten state can take place on the leading edges of tubes and tube plates which are suitably situated with respect to the gas flow and temperature, and will spread as a sticky film over the surface, thus trapping more solid particles.[4] These latter may not have been present in the crude oil but can be the result of mal-operation of catalytic crackers, which feature in modern refineries. They are known as 'cat. fines' and consist of alumino-silicates which form the carrier for the catalyst. Their occurrence however is accidental rather than normal. They serve to increase the bulk of the deposits rather than to cause them. Table 10.2 shows an oil ash analysis compared with that for coal A from Table 10.1.

It will be seen that there is a great difference between these two analyses, the major ash constituents in oil being vanadium followed by nickel and sodium. The first two are probably organically combined in the oil, whilst the sodium can be present from sea-water contamination or from refinery processes.[4] The vanadium normally burns to the pentoxide $V_2O_5$ in most cases, but with very low excess air, less than say 3%, which is unusual for the types of boiler discussed in this book, the tetroxide, $V_2O_4$, may be formed. The former has quite a low melting point, about 800 °C, a temperature which is usually exceeded at the entry to the convection passes

**10.2**  Analyses of coal and oil ashes

| Coal, ash content 7.5% | | Oil, ash content 0.1% | |
| --- | --- | --- | --- |
| *Element* | *% in ash* | *Element* | *% in ash* |
| Si | 40 | Si | 3.6 |
| Al | 26.2 | Al | 0.3 |
| Fe | 20.0 | Fe | 1.8 |
| Mg | 1.5 | V | 68.9 |
| Ca | 3.5 | Ni | 17.1 |
| Ti | 1.1 | Pb | 0.9 |
| Na | 2.9 | Na | 7.4 |
| K | 4.8 | Cu, Ca, Sn | 0.1 each |

of most boilers at full load. Vanadium pentoxide, as such and in a molten state is probably present in these gases and can deposit on cooler tubes. It may then solidify but will react with other substances, particularly sodium to form a variety of sodium vanadates, e.g. sodium metavanadate $Na_2OV_2O_5$ with a melting point of 621 °C. There is great variety and complexity in the compounds of vanadium, sulphur, iron, sodium and nickel that can be formed and may have quite low melting points, often sufficiently low to bond other particles to a deposit that has already formed.

### 10.2.2   Control of deposits from oil firing

It is clear that the lower the gas temperature the less likely are heavy deposits to form. The larger the furnace of a boiler, the lower will be the temperature of the gas entering the superheater and convective zone (Ch. 5). Thus deposit problems will be less with a conservatively designed boiler than with one of more compact design.

Over the last few years oil refining practice has changed to enable more of the lighter fractions to be obtained from the crude oil for the premium market: transport. Thus the residual boiler fuel is likely to contain more of the deposit-forming substances than was formerly the case. These fuels are sometimes burned in types of boiler not specifically intended for their use, with a result that severe deposition can occur. An example of a case in which this has occurred is in the closely spaced convection surfaces of relatively highly rated shop-assembled boilers which have very restricted access for cleaning and lack spaces into which deposits can fall to prevent restriction of the gas flow. In such cases gas temperatures and tube arrangements cannot be altered, except that severe de-rating can reduce the former, but this is not often acceptable. This leaves only one channel open for amelioration, i.e. chemical modification.

Magnesia, besides effectively suppressing the formation of the initial sodium pyrosulphate layer, will combine with the vanadates and raise their melting points considerably. The overall effect is to produce deposits which are light and friable compared with those not so treated, and which can easily be removed by sootblowing. For instance, magnesium vanadate 3 $MgO.V_2O_5$ has a melting point of 1243 °C compared with 621 °C for $Na_2O.V_2O_5$. A detailed account of the effects of magnesia addition to the furnaces of two highly rated water-tube boilers has recently been given[5].

## 10.3   Deposits from refuse- and waste-derived fuels

Other fuels can give deposit problems; currently there is wide interest in the burning of refuse-derived fuels (RDF). Two requirements can be met by this process: the reduction in volume and the sterilisation of the refuse, and the reclamation of useful heat. Refuse can contain a wide variety of substances which can cause deposition, including iron, aluminium, and glass. There is however very little sulphur, so that the reactions involving this substance, so important with coal or oil, are unlikely to occur. Boiler deposits, however, take place to a serious extent; the mechanisms involved and methods of control are currently under close study by ETSU (Energy Technical Support Unit of the United Kingdom Atomic Energy Authority). Refuse is also likely to contain chlorinated plastics; there is thus a potential risk, due to the release of chlorine compounds and their interaction with other substances, of creating complications with both deposits and corrosion.

The majority of fuels derived from biomass waste do not contain much ash, but they are rich in potassium which is a strong flux. Some vegetable wastes such as rice husks contain silica, which at high temperatures reacts readily with potassium compounds to form a glass so that deposition from this cause may be likely. Potassium is also highly aggressive towards refractory materials containing silica, and furnaces for burning such wastes should therefore contain a minimum of refractory material; what there is should be of a high alumina rather than high silica content.

## 10.4   Corrosion

Two types of fire-side corrosion can occur in boilers:

1.   high-temperature corrosion; and
2.   low-temperature corrosion.

The first of these is rarely encountered except in high-pressure boilers used for power generation where superheater metal temperature can approach 600 °C, and the boilers associated with the mass burning of municipal refuse and refuse-derived fuels. It may not be entirely absent however, so that the problem will be discussed briefly. Low-temperature corrosion is however quite common in economisers, air preheaters, chimneys and ducting, and therefore will be discussed in greater depth.

### 10.4.1   High-temperature corrosion

This is mentioned first because it is associated with the foregoing section on deposits. It will be realised that the bonding salt, sodium pyrosulphate, will be present in most cases. This substance is highly corrosive to steel, and if allowed to persist will cause corrosion to take place beneath the deposits. In general, molten salts will corrode. The various vanadates present in deposits arising from oil ash are also corrosive whilst molten. The problem seems to be therefore to avoid the molten state which, it is shown in the previous section, can be done by careful design or by the use of magnesia; the latter will also neutralise any $SO_3$ produced by partial decomposition of the salts.

High-temperature corrosion has been a feature of boilers associated with mass burning refuse incinerators, particularly if there are heating surfaces exposed to the gases in the combustion chamber or high metal temperatures. It is believed that the high hydrogen chloride (HCl) content of the gases is a major cause of this. This HCl results from the high chlorine content of the plastics contained in the refuse. The situation is aggravated by reducing conditions which exist due to the heterogeneous nature of the fuel, the deep fuel beds used and hence the difficulty in achieving the correct air-to-fuel ratio at all points in the furnace.

The current remedy is to prevent the gases coming into contact with the tubes until combustion is complete and for the gases to be thoroughly mixed by the use of at least 100% excess air and by the injection of secondary air. This is done by either covering the tubes with refractory material or by making the furnace independent of the boiler. Even after taking these precautions it is necessary to avoid tube metal temperatures above about 425 °C if high-temperature corrosion of superheaters is to be avoided. This limits the steam conditions that can be used with incinerators if high availability is required.

The result of high-temperature corrosion is to reduce the outside diameter of the tubes and, quite often, to flatten one face, both of these processes being detected during off-load inspection (see Fig. 10.2).

### 10.4.2 Low-temperature corrosion

#### 10.4.2.1 Occurrence

This is a far more common problem than is high-temperature corrosion, and it can occur in all types and sizes of boiler where sulphur-bearing fuels are burned and where temperature conditions are favourable to the condensation of sulphuric acid. The sulphur in the fuel burns mainly to sulphur dioxide.

$$S + O_2 \longrightarrow SO_2 \tag{10.2}$$

If excess oxygen is present, and it always is in practical cases, further oxidation takes place.

$$SO_2 + \tfrac{1}{2}O_2 \longrightarrow SO_3 \tag{10.3}$$

This reaction can be catalysed by the presence of hot metallic oxides of iron, both in the boiler construction and the fuel ash, and, with fuel oil, of vanadium and nickel. The greater the amount of excess air, the greater will

Outer corroded
tube surface

**10.2** Cross section of a corroded furnace tube from a water-tube boiler

be the amount of $SO_3$ produced. This latter fact is used in the operation of large water-tube boiler plant to control $SO_3$, i.e. minimising oxygen content will minimise $SO_3$ production, so in such plant excess oxygen is controlled to about 0.5% or even less. This also benefits plant efficiency, but necessitates accurate burner manufacture and sophisticated oil fuel ratio control, both of which raise the boiler cost.

The hydrogen in the fuel oxidises to water vapour.

$$2H_2 + O_2 \longrightarrow 2H_2O \qquad\qquad [10.4]$$

In the cooler parts of the boiler this reacts with the $SO_3$ to form sulphuric acid.

$$H_2O + SO_3 \longrightarrow H_2SO_4 \qquad\qquad [10.5]$$

Sulphuric acid has a high boiling, and therefore condensation, point of 290 °C for the concentrated form, less for diluted acid. Under boiler conditions this is generally in the range 125 to 145 °C. The temperature of condensation (the acid dewpoint) is an important factor in boiler operation, as surfaces below this figure will condense acid and be subject to corrosion. The lower the temperature of the surface the greater, generally, will be the amount of corrosion.

Figure 10.3 illustrates typically the relationship between surface temperature and corrosion rate. It will be seen that after reaching a peak at 20–30 °C below the acid dewpoint, the corrosion rate decreases. This is due to the formation of acid mist in a cool environment, which carries the acid out of the system. A further reduction in temperature, however, causes a rapid increase in corrosion rate, which, at low temperatures, can be catastrophic. This arises from the increased volume of acid produced due to dilution as the water dewpoint is approached, the volume being

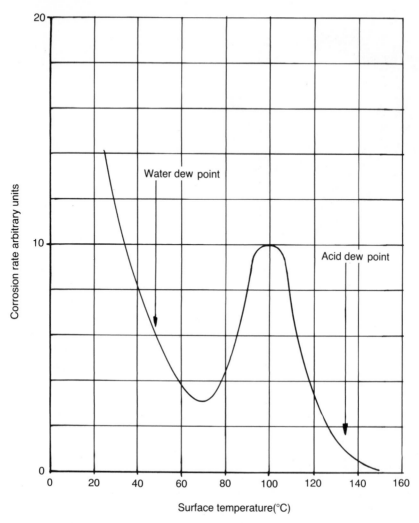

**10.3**   A typical relationship between surface temperature and corrosion rate

sufficient to leach any products of corrosion which would otherwise form a partly protective layer. Dilute acid is also more reactive than is concentrated acid. The acid dewpoint can be measured using the apparatus described by Land.[6] This consists of a glass surface, cooled to any required extent by a blast of air on the rear surface, which is not in contact with the gases. A thermocouple is sealed into the surfaces and a ring of platinum surrounds this. An e.m.f. of about 11 V a.c. is applied between the thermocouple and the ring, and the air blast is adjusted until the current flow is steady at 5 $\mu$A. The thermocouple reading indicates the dewpoint which corresponds with the initial deposition of electrically conducting acid. This apparatus is widely used and is commercially available.

The more $SO_3$ that is formed, the higher will be the acid dewpoint, but this is also affected by the amount of water vapour present. Pierce[7] quotes a transposition of an equation originally proposed by Verhoff and Banchero.[8]

$$\frac{1000}{T} = 1.7846 + 0.0267 \log_{10} P_1 - 0.1031 \log_{10} P_2 + 0.0329 \log_{10} P_1 \log_{10} P_2$$
$$[10.6]$$

Where: $T$   is acid dewpoint temperature (K);
   $P_1$   is partial pressure of $H_2O$ (bar); and
   $P_2$   is partial pressure of $SO_3$ (bar)
   This is said to be accurate within $\pm$ 7 K.

If the $SO_3$ in the gases can be estimated, this correlation may be used to estimate the dewpoint. The reverse, estimation of $SO_3$ from dewpoint, is not true, however. This is due to the possibility of basic substances, coal ash or neutralising additives being deposited on the dewpoint meter and affecting its reading. Dewpoint can be a true indication of the vulnerability of a surface to corrosion but not of the $SO_3$ content of the bulk gases flowing over the dewpoint probe. Dewpoint measurement should not be used as an indication of $SO_3$ content of the flue gases. This should, if required, be done by a chemical measurement such as that described by Corbett.[9]

### 10.4.2.2   Control of low-temperature corrosion

It will be clear that all surfaces in contact with the flue gases in a boiler plant will be liable to corrosion if their temperatures are below the acid dewpoint of the gases. It follows that their temperatures should be kept above the acid dewpoint, but this is not always possible. Examples are given below.

1.   In low-temperature hot-water boilers where the water temperature is below, say, 100 °C heavy fuel oil should not be used since the acid dewpoint will be at least 125 °C. The alternative is to use gas oil which, with its lower sulphur and catalyst content, will give a dewpoint of around 80 °C. Coal with a sulphur content of around 1.5% will give a dewpoint probably in the range 105 °C to 115 °C, but the effects of this are largely negated by the fly ash produced, some of which will settle out on the surfaces. This will dry the acid off and, since it is calcined and therefore partly basic, will tend to neutralise the acid, albeit with the need for frequent cleaning of the surfaces to remove the deposits. Natural gas, being virtually sulphur free, does not usually present the problem.

2.   In economisers the temperature of the water entering the system is usually below the acid dewpoint, in which case corrosion can be expected. The condition may be alleviated by recirculating water from

the boiler or the economiser outlet to give a mixed water temperature above the dewpoint, although this will increase the heating surface and hence size of economiser to achieve a given outlet gas temperature and, hence, efficiency (section 7.4.1.2).

3.  In air preheaters the incoming air will normally be at ambient temperature; therefore condensation will take place at the inlet end, and, unless the heater is carefully designed (section 7.4.2.4), so will corrosion. Some alleviation can be obtained by air recirculation, in a similar manner to that with water recirculation with economisers or by preheating the entering air with a steam or hot-water preheater. The steam or hot water can be taken from the boiler or from an external source of waste heat. A small advantage can be obtained by ducting the air intake from the top of the boilerhouse, where temperatures are higher. Tubular air heaters are sometimes made from glass, which is not attacked by sulphuric acid. (section 7.4.2.4).

4.  In ducting, chimneys and grit arrestors, all surfaces and particularly flanged joints should be well insulated externally to maintain the internal surfaces at as high a temperature as the temperature of the flue gases will permit. In the case of very highly efficient plant using air preheaters the flue gas temperature may not be sufficient for this purpose. In such cases the use of a neutralising agent such as magnesium hydroxide may be essential.

5.  In intermittently operating plant, depending on the extent of idleness, surfaces may cool sufficiently to allow condensation and corrosion to occur. This is often tolerated, but should corrosion become a serious problem the use of a neutralising agent immediately before shutting down and on start-up is recommended. Plant idle for long periods should be cleaned as soon as possible after shut-down and all surfaces given a dusting with a neutralant. Water washing and treating with an alkaline solution is also acceptable, but the acidic wash-water effluent will require neutralisation before disposal to drain. During prolonged shut-down the whole boiler can be subjected to corrosion, since ferrous parts that are normally too hot for this to occur, will become cold and, if in contact with acidic products, these will absorb moisture and set up corrosion.

Where the various measures discussed above cannot be employed the remaining method of alleviating corrosion is to use a neutralising agent such as $Mg(OH)_2$. In sections 10.1.2.5 and 10.2.2 of this chapter the use of airborne magnesia injected into the furnace, was discussed in relation to deposits. For control of corrosion it should be injected as an airborne suspension immediately upstream of the parts subject to corrosion, e.g. upstream of the economiser. The quantity injected should be proportional to the amount of $SO_3$ likely to be formed, which is normally in the range 10–30 p.p.m. of flue gases. The reaction is

$$H_2SO_4 + Mg(OH)_2 = MgSO_4 + 2 H_2O \qquad [10.7]$$

The product, $MgSO_4$, is non-toxic and readily removed from the surface by soot blowing or water washing. The reaction takes place mainly on the surfaces on which the acid condenses, but there is some evidence to show that there is some gas-phase reaction as well.

Attempts have been made to use ammonia gas as a neutralant, but complications arise from two compounds being formed when the mixing with flue gas is non-uniform: the normal ammonium sulphate, which is innocuous, and the bisulphate, which forms deposits which are acidic and therefore can cause corrosion. The latter compound forms when the amount of ammonia present is insufficient to form the normal sulphate.

The phenomenon of smutting is associated with corrosion (see section 7.3), being due to the accretion of soot particles on acid condensed on cool walls. In neutralising and drying this acid off, magnesia reduces smutting. It will be seen that magnesia has a beneficial effect in controlling deposits, corrosion and smutting, but it is stressed that this, or any other neutralising agent and the storage and injection equipment is expensive, and should only be used when necessary. It will also be seen that there is no mystery about its action: its effect is simple chemistry.

# References

1. *Report on View of Available Information on Corrosion and Deposits in Coal Fired boilers and Gas Turbines to the American Society of Mechanical Engineers.* (Undated) Battelle Memorial Institute.
2. Crumley, P. H. and Fletcher, A. W. 'The formation of sulphur trioxide in flue gases', *J. Inst. Fuel* 1956 **29:** pp. 322–7.
3. *Coal Fouling and Slagging Parameters.* E.C. Winegartner ASME, 1947.
4. Crossley, H. E. 'External fouling and corrosion of boiler plant; a commentary', *J. Inst. Fuel* 1967 (Aug.) p. 342.
5. Bell, B., Dewey, A. and Turner, J. Institute of Energy Seminar, Ossett November 1986.
6. Land, T. J. *Inst. Fuel* 1977 **50:** pp. 68–75.
7. Pierce, R. R. 'Estimating acid dewpoints in stack gases', *Chem. Eng.* 1977 (Apr.) p. 11.
8. Verhoff, F. H. and Banchero, J. T. 'Predicting dewpoint of flue gases', *Chem. Eng. Prog.* 1974 **70:** p. 71.
9. Corbett. P. F. 'The determination of $SO_2$ and $SO_3$ in flue gases', *J. Inst. Fuel* 1951 (Nov.) p. 247.

# CHAPTER 11

# Thermal testing

## 11.0  Introduction

It is usual for a boiler manufacturer to prove to the purchaser that the boiler, after commissioning, can achieve its rated output and stated efficiency; this is an 'acceptance' test and is carried out very formally as legal matters may be involved if the boiler does not meet, within stated limits, its guaranteed output and efficiency. Once the boiler has been accepted, its efficiency (and output) may be influenced by various factors such as maladjustment of the controls, fouling, or a change from specified operating conditions. Both maximum output capability and efficiency should therefore continue to be measured during its working life, or at least those factors which affect efficiency, mainly exit gas composition and temperature, should be. It is the purpose of this chapter to explain how efficiency can be measured.

## 11.1  Boiler efficiency

The thermal efficiency of a boiler is the ratio of useful energy output to the energy input. By far the greatest component of the latter is the energy in the fuel; that supplied as power to the auxiliaries being negligible in comparison. Efficiency is expressed as the percentage of the energy input which appears as useful output, and is always less than 100%; 80–90%, based on the gross calorific value, is typical of modern plant. The difference between energy input and output is the sum of the various energy losses from the boiler. These are:

1.  the flue gas loss (generally the most important);
2.  the loss due to unburned fuel;

3.  the loss in hot water blown down from the boiler;
4.  the losses due to radiation, convection and conduction from the boiler structure; and
5.  sensible heat in the ash of very high-ash fuels.

The efficiency of a boiler can therefore be expressed alternatively as 100% minus the sum of the losses expressed as a percentage of the input energy.

## 11.2   The measurement of boiler efficiency: standards

The two definitions of efficiency given above lead to two methods of measuring it:

1.  by measuring input and output (this is called the 'direct method'); and
2.  by measuring individual losses, totalling them and deducting the sum from 100%. (This is called the 'indirect method', but the terms 'losses method' and 'efficiency by difference' are also used.)

Ideally, both methods should be used, one as a check upon the other, and a complete heat balance obtained. There will be a difference between the results obtained from each, this difference being due to human and instrumental errors. Testing by both methods is rarely done, however, the costs of doing so being prohibitive compared with the benefits obtained. Having obtained two values for efficiency, the question would arise which one should be accepted in case of dispute.

It is now generally accepted[1,2,3] that the indirect method is the simpler to carry out and yields the more accurate results. It also identifies the loss areas which need attention should a shortfall in efficiency below the expected value occur. The reason for the greater accuracy of the indirect method is really quite simple. Whereas the direct method needs to measure the output, which is around 80% of the input, the indirect method only measures 20% of the input. A one per cent error in the direct method therefore gives an error of 0.8 efficiency points, whereas a similar error in the indirect method only gives an error of 0.2 efficiency points. The full analysis of the probability of errors occurring is much more complex than this, but the basic principle is similar. BS845 'Acceptance tests for industrial type boilers and steam generators' dealt with the complete heat balance; it has now been revised in two parts: Part One dealing with more or less standard boilers, where the indirect method is specified; and Part Two, which is concerned with complex plant where there are many channels of heat flow. In this case, both the direct and indirect methods are applicable, in whole or in part. With both methods the overall tolerance permitted is plus or minus two per cent, but in most cases the measurement of efficiency is likely to be well within this provided that the requirements laid down in the Standard are satisfied. The equivalent USA Standard is 'ASME PTC-4-1 Power Test Code Steam Generating Units'. This includes both the direct and indirect methods and is comparable with BS2885:1974 'Acceptance tests on stationary steam generators of the power station type' and with Part Two of the revision of BS845.

Where thermal testing is conducted for contractual purposes, it is essential that the agreed standard be implemented rigorously and properly witnessed; where routine tests are involved the requirements are less stringent but in his own interests the boiler user is advised to ensure that accurate instruments in good working condition are used in accordance with the methods described in the Standard. Testing standards tend to be long and detailed and it is not proposed to describe them in this chapter, but rather to outline the principles involved.

### 11.2.1 The direct method

#### 11.2.1.1 Heat input

Both heat input and heat output must be measured. The measurement of heat input requires a knowledge of the calorific value of the fuel and its flow rate in terms of mass or volume, according to the nature of the fuel. With natural gas the process is simple: a gas meter of the type approved for the sale of gas is used and the measured volume is corrected for temperature and pressure. A sample of gas can be obtained for calorific value determination, but it is usually acceptable to use the calorific value declared by the gas suppliers. It is strongly advised that approved gas meters be permanently installed in every boiler house using this fuel.

Oil can be dealt with in the same way, but it is rather more difficult. Heavy fuel oil is very viscous, and this property varies sharply with temperature. The meter, which is usually installed on the combustion appliance, should be regarded as a rough indicator only and, for test purposes, a meter calibrated for the particular oil to be used and over a realistic range of temperature should be installed. Even better is the use of an accurately calibrated day tank.

The accurate measurement of the flow of coal or other solid fuel is very difficult, there is no short cut, measurement must be by mass, which means that bulky apparatus must be set up on the boiler-house floor, and the coal manhandled. Samples must be taken and bagged throughout the test, the bags sealed and sent to a laboratory for analysis and calorific value determination. In some more recent boiler houses, the problem has been alleviated by mounting the hoppers over the boilers on calibrated load cells, but these are as yet uncommon.

Continuous coal flow measurement for the assessment of fuel

consumption and boiler efficiency during normal operation of stoker-fired boilers is achieved with a reasonable degree of accuracy by continuous measurement of fuel bed depth and grate speed. These factors, together with the known width of the grate, enable the volume of coal used to be continuously calculated by a microprocessor. The fuel bulk density is input to the microprocessor using information obtained from measurements of the density of samples taken at intervals. The microprocessor can then indicate and record the ongoing fuel consumption in suitable units.

### 11.2.1.2 Heat output

There are several methods which can be used for measuring heat output. With steam boilers, an installed steam meter can be used to measure flow rate, but this must be corrected for temperature and pressure. In earlier years, this approach was not favoured due to the change in accuracy of orifice or venturi meters with flow rate. It is now more viable with modern flow meters of the variable-orifice or vortex-shedding types. It is not usually easy to install a meter specially for a test, as bends in pipes can affect its accuracy. The alternative with small boilers is to measure feedwater, and this can be done by previously calibrating the feed tank using weighed increments of water to fill the tank from a marked low level to a marked high level, and operating the tank between these limits. The number of fills are counted and, finally, the intermediate position is interpolated when the test ends. Where feedwater is measured, however, it is important to allow for boiler blowdown and, with saturated steam boilers, steam wetness.[4] Normally, this latter should not exceed about 2% of the output, but it can be much more than this if water conditions are liable to cause foaming (see Ch. 6).

With hot-water boilers the heat output is measured by an installed water meter, preferably of the variable-orifice type. The temperatures of the water entering and leaving the boiler are also required. With low-temperature hot-water systems, of which there are many, the difference between flow and return water temperatures can be as little as 20 °C, in which case an error of only 1 °C in the measurement of this differential is equivalent to an error of 5% in the measurement of heat output!

### 11.2.2 The indirect method

The efficiency can be measured quickly and easily by measuring the losses using the principles to be described. Where, as in an acceptance test, the output is needed, this must be measured as described in section 11.2.1.2 or by multiplying the input, measured as described in section 11.2.1.1, by the efficiency. In some cases, e.g. with waste-heat boilers (see section 16.1), the input can be difficult to measure, in which case the output must be used and its possible inaccuracies accepted. Dividing this by the efficiency makes an estimate of input possible. This can be very useful in itself, for instance in determining the calorific value of a waste fuel. The important part of the indirect method, however, is the measurement of the losses. This is dealt with in detail in BS845:1987 Parts One and Two, but the formulae are summarised here.

The flue gas loss is divided into two parts: the sensible heat in the gases leaving the boiler, excluding water vapour; and, when calculating the efficiency as a proportion of the *gross calorific value* of the fuel, the sensible and latent heat in the water vapour. The sensible heat in the gases can be obtained from first principles: for each constituent the mass multiplied by its specific heat and by the temperature difference between the exit gas temperature and the ambient temperature. Siegert[5], however, developed simplified formulae for the 'dry gas' and 'water vapour' losses which were later modified and improved by Dunningham and Nonhebel.[6] These are now incorporated into BS845 and are very simple to use, taking into account the net amount of fuel actually burned. The loss formulae giving losses as a percentage of the gross calorific value are:

$$\text{dry gas loss } L_1 = \frac{k(t_3 - t_a)(1 - 0.01(L_4 + L_5))}{V_{CO_2}} \qquad [11.1]$$

loss due to water vapour

$$L_2 = \frac{(m_{H_2O} + 9H)(2488 - 4.2t_a + 2.1t_3)}{Q} \qquad [11.2]$$

loss due to unburned gases

$$L_3 = \frac{k_1 V_{CO}(1 - 0.01(L_4 + L_5))}{V_{CO_2} + V_{CO}} \qquad [11.3]$$

loss due to unburned carbon in ash and riddlings

$$L_4 = \frac{33\,820\,M_1\,a_1}{M_f\,Q} \qquad [11.4]$$

loss due to unburned carbon in grit and dust

$$L_5 = \frac{33\,820\,M_2\,a_2}{M_f\,Q} \qquad [11.5]$$

Where: 
$k$      is the Siegert constant (see section 11.2.3);
$t_3$      is the temperature of the exit gas from the boiler (°C);
$t_a$      is the temperature of the air entering the boiler (°C);
$V_{CO_2}$      is the percentage $CO_2$ in the dry flue gases (% v/v);
$m_{H_2O}$      is the percentage moisture in the fuel as fired (% m/m);
$H$      is the percentage hydrogen in the fuel as fired (% m/m);
$Q$      is the gross calorific value of the fuel as fired (kJ kg$^{-1}$);
$V_{CO}$      is the percentage CO in the exit gases (% v/v);
$k_1$      is a constant (see section 11.2.3);
$M_1$      is the mass of ash and riddlings collected during the test (kg);

$M_2$   is the mass of grit and dust collected during test (kg);
$M_f$   is the mass of fuel fired during test (kg);
$a_1$   is the percentage carbon in ash and riddlings (% m/m); and
$a_2$   is the percentage carbon in grit and dust (% m/m).

The loss by radiation, convection and conduction from the boiler enclosure is, with modern boilers, a small percentage, varying from 1.0% for firetube and water-tube boilers up to 2 MW output, down to 0.3% for boilers above 5 MW. These values, together with a formula for boilers of unusual geometry, are given in BS845:1987.

## 11.2.3  Notes on the formulae

### 11.2.3.1  Equation [11.1]

The constant $k$ is known as the Siegert constant. Its value depends on the carbon content and the calorific value of the fuel as fired

$$k = \frac{255C}{Q} \tag{11.6}$$

Where: $C$   is the carbon content of the fuel (% m/m); and
$Q$   is as defined in section 11.2.2.

Typical values of $k$ are:

Coal            0.62
Heavy fuel oil  0.51
Gas oil         0.48
Natural gas     0.35

It will be seen that the loss $L_3$ due to unburned gases has been omitted from the unburned losses in equation [11.1]. This is because the quantity should be negligible if the combustion appliance has been correctly adjusted.

Where oxygen rather than $CO_2$ is measured, as is often the case, the conversion is

$$V_{CO_2} = \left(1 - \frac{V_{O_2}}{21}\right)\bar{V}_{CO_2}$$

Where: $V_{O_2}$   is the percentage oxygen in the dry gases (% v/v); and
$\bar{V}_{CO_2}$   is the stoichiometric (theoretical) carbon dioxide in the dry flue gases (% v/v).
Typical values of $\bar{V}_{CO_2}$ are:

Coal            18.4%
Heavy fuel oil  16.0%
Gas oil         15.6%
Natural gas     11.8%                        (See Fig. 3.1)

### 11.2.3.2  Equation [11.2]

Typical values for $H$ are:

| | |
|---|---|
| Coal | 4.1% m/m |
| Heavy fuel oil | 11.4% m/m |
| Gas oil | 13.0% m/m |
| Natural gas | 24.4% m/m |

The value of $m_{H_2O}$ is very significant for coal and may vary over a range 5–20%. It is important therefore that this quantity be determined as accurately as possible.

### 11.2.3.3  Equation [11.3]

$L_3$ should be thermally insignificant. If more than a few hundred parts per million of CO are present, unacceptable smoke will be emitted, except with gasfiring. Sensitive apparatus, infra-red or colorimetric, will be needed to measure small concentrations of CO, and the combustion appliance should be adjusted until a satisfactorily low reading, say 200 p.p.m. is achieved. The test should otherwise be abandoned until correction is made. Typical values for $k_1$ are:

| | |
|---|---|
| Coal | 63 |
| Oil | 53 |
| Natural gas | 40 |

### 11.2.3.4  Equations [11.4] and [11.5]

These apply mainly to solid-fuel firing, although equation [11.5] can be significant when firing heavy fuel oil. In this case, however, there is a statutory limit which, depending on boiler size, limits emission to 0.4% or less of the fuel fired. It is generally acceptable, therefore, to allow this thermally small loss without requiring measurement, which can be difficult to arrange.

## 11.3  Test procedure

This chapter is not intended to detail the full procedure for a boiler test. The Standards, particularly BS845:1987, should be consulted and their requirements implemented. BS845 lists other British Standards which are relevant to boiler testing, and these should be consulted and implemented as necessary and be determined by the purpose of the test and the type of fuel used. For instance, if coal is used it is important that BS1017:Part One: 1977 (= ISO1968) 'Sampling of coal' be studied and applied.

It will have been noted that gas temperature and composition occur in equations [11.1], [11.2] and [11.3]. It is important that the measurements are representative of the bulk of the gas. It is suggested that the exit duct of the boiler be probed and traversed to find the location of the zone of maximum temperature; this is likely to coincide with the zone of maximum gas flow and is therefore a good sampling point for both temperature and gas composition.

It will be found to be convenient if the gas sampling tube is combined with the temperature measuring thermocouple, thus ensuring that the gas temperature and composition sampling points are the same. The thermocouple should be of small-diameter bare wires exposed to the gas flow. This is to minimise heat loss from the tip, which tends to give a low reading. At boiler exit temperatures in the range 150–250 °C the error is small, but if it is desired to examine the heat flow in the hotter parts of the boiler where the gas temperature may be 800 °C or more, serious errors can occur.[7] In such cases the use of a suction pyrometer is recommended.

## 11.4   Numerical example

The following data evolved from a test on a coal-fired boiler rated at 5.5 MW output.

| | |
|---|---|
| Coal fired ($M_f$) | 1000 kg hr$^{-1}$ |
| Ash collected ($M_1$) | 100 kg hr$^{-1}$ |
| Grit collected ($M_2$) | 30 kg hr$^{-1}$ |
| Carbon in ash ($a_1$) | 20% |
| Carbon in grit ($a_2$) | 35% |
| Calorific value of fuel as fired ($Q$) | 26 000 kJ kg$^{-1}$ gross |
| Moisture content of fuel as fired ($m_{H_2O}$) | 16% |
| Carbon content of fuel as fired ($C$) | 65% |
| Hydrogen content of fuel as fired ($H$) | 4.1% |
| $CO_2$ content of exit gases, dry basis ($V_{CO_2}$) | 12.0% |
| CO content of exit gases, dry basis ($V_{CO}$) | 200 p.p.m. (negligible) |
| Temperature of exit gases ($t_3$) | 230 °C |
| Temperature of air entering boiler ($t_a$) | 20 °C |

Calculate the loss items described in section 11.2.2, the thermal efficiency of the boiler based on the gross calorific value of the fuel, and the thermal output of the boiler.

First calculate the unburned carbon losses $L_4$ and $L_5$ using equations [11.4] and [11.5].

$$L_4 = \frac{33\ 820 \times 100 \times 20}{1000 \times 26\ 000} = 2.60\%$$

$$L_5 = \frac{33\ 820 \times 30 \times 35}{1000 \times 26\ 000} = 1.36\%$$

From equation [11.6] calculate $k$.

$$k = \frac{255 \times 65}{26\ 000} = 0.64$$

From equation [11.1] calculate $L_1$.

$$L_1 = \frac{0.64 \times (230 - 20)(1 - 0.01\ (2.6 + 1.36))}{12} = 10.75\%$$

From equation 11.2 calculate $L_2$.

$$L_2 = \frac{(16 + (9 \times 4.1))(2488 - (4.1 \times 20) + (2.1 \times 230))}{26\ 000} = 5.88\%$$

For a 5.5 MW boiler the loss due to radiation, etc. is

|  |  |
|---|---|
|  | 0.30% |
| Total losses | = 20.89% |

Efficiency = 100% − total losses  = 79.11%
Output = input × efficiency

$$= \frac{1000 \times 26\ 000 \times 79.11}{100}$$

$$= 20\ 568\ 600 \text{ kJ hr}^{-1}$$

$$= \frac{20\ 568\ 600}{3600} = 5713 \text{ kW}$$

$$= 5.7 \text{ MW}$$

# References

1. Brown, F. L. and Murray, M. V. British Coal Utilisation Research Association Information Circular No. 86, 1953.
2. Doxey, G. 'Some notes on the routine determination of boiler heat balances', *J. Inst. Fuel* 1956 (Jan.) p. 11.
3. Downie, J. M. 'Accurate', *Maintenance Engineering 1974* (Sept.) p. 58.
4. BS 3812:1964 *Estimating the Dryness of Saturated Steam*. British Standards Institution, London.
5. Siegert, A. 'On the determination of heat loss in chimney gases' *J. Gasbeleuchtung* 1888 **12**: p. 736.
6. Dunningham, A. C. and Nonhebel, G 'The calculation of periodic boiler plant efficiencies', *J. Inst. Fuel* 1953 (Mar.) p. 387.
7. McAdams, W. H. *Heat Transmission* (3rd edn). McGraw-Hill. Ch. 10: pp. 261–4.

# CHAPTER 12

# Firetube boilers – design for output, efficiency and reliability

## 12.0   Introduction

The purpose of a boiler is to raise steam, or to heat water at specified conditions of output, temperature and pressure. The main operational cost with fossil-fuel (coal, oil and natural gas) fired boilers is normally that of the fuel consumed; thermal efficiency is therefore important. The provision of steam, for power or process, or hot water for process and space heating, is vital to the viability of the premises which boilers serve, so that uninterrupted supplies of the commodities are required. The intention of this chapter is to indicate how the ability to provide the specified output, high efficiency, and reliability are incorporated into a boiler at the design stage.

It should be pointed out that in the interests of economy and delivery, most boilermakers have established standard ranges of boilers, and in many instances therefore it is a question of selecting a suitable boiler for a project from the standard products available rather than designing one for each project. These basic requirements will have been incorporated into the original design.

## 12.1   Output and operating conditions

These will be specified by the prospective user. With both hot-water and steam boilers the thermal output is basic. This is expressed in terms of kW (formerly BTU $hr^{-1}$), but pressure and steam temperature are also important, particularly where the steam is used for generating power, in which case relatively high pressures and some superheat are required. Many hot-water boilers used for large-scale space heating, e.g. district heating, need a high water temperature to convey the heat generated by the boiler economically and effectively. To attain these high temperatures high pressures are needed.

The release of steam in a firetube boiler requires a certain area of water surface from which to disengage the steam, and a certain height above the water plane to allow the water level to rise with increasing load and still allow sufficient area to disengage the steam without carry over of water taking place (Ch. 6). It will be appreciated that, as the water level rises with increasing load (due to the presence of more steam below the water level in the boiler), the steam disengagement area will decrease because, with the water level being above the centreline of the boiler, the sides of the containing shell converge. It is recommended that the area at the normal water level (NWL) should be such that the steam released will do so at a velocity not exceeding $0.056$ ms$^{-1}$ and that the minimum height of the steam offtake above the NWL should be greater than 20% of the shell diameter. These data should be regarded as good practice; they are not contained in any constructional standard. They do, however, have a major influence on the length and diameter of the shell, as do the accommodation of the heat-transfer surfaces and the firing appliance. Hot-water boilers, unless steam pressurised (which is now rare), are fully flooded, no steam space being needed.

The heat-transfer surfaces comprise the furnace, the reversal chamber or combustion chamber, and the convection tubes. The selection of the dimensions of these will now be considered.

## 12.2 The furnace

This, in a firetube boiler, is nearly always cylindrical in shape, although for special purposes, e.g. the accommodation of some types of fluidised beds, it may be quite different. It accommodates the firing appliance and, along with wet-back reversal chamber surfaces, transfers sufficient heat from the burning fuel to cool the gases to a temperature acceptable to the tubular convection pass. With dry-back boilers, the reversal chamber is refractory lined and transfers little significant heat. The furnace of a dry-back boiler therefore needs to provide more heat-transfer surface than that of a wet-back boiler.

Gas, oil or pulverised fuel burners are quite easily accommodated at the front of the furnace, one burner being used per furnace. Grates for the combustion of solid fuel, or fluidised beds, require appreciable space, and screen off much of the surface otherwise available for heat transfer. The furnaces of boilers fired by solid fuel are therefore considerably larger than those needed for gas or oil. Pulverised fuel and coal/water mixture (CWM) burners occupy an intermediate position governed by the lower combustion rate obtainable with these fuels compared with that with gas or oil.

All fuels need time in which to burn, and combustion must be complete in the furnace. This is a requirement of BS2790:1986. To accomplish this, a certain furnace volume and proportions will be needed. With oil and gas firing, an appropriate figure for volumetric heat release used in the UK is up to 1.8 MW m$^{-3}$ [1], but for coal firing on a stoker much less will be acceptable because the mixing of combustible gases and air over the fuel bed will be less well defined and less intense. A figure of 0.8 MW m$^{-3}$ is typical, although higher values are sometimes used, albeit often with associated problems. These figures are based on the net furnace volumes only, excluding the volume of the reversal chamber, and the volume occupied by burner refractory or firing appliances which, with coal firing in particular, can be very appreciable, as has been mentioned earlier. In the USA, a slightly different criterion is used, namely 1.55 MW m$^{-3}$, but in this case the geometric projection of the furnace dimensions into the reversal chamber is allowed, which makes the figure about the same as in the UK based upon the actual furnace dimensions. This is for gas or oil firing; no figure is quoted for coal firing.

The proportions of the furnace are important and should suit the flame pattern of the burner used. In practice, the range of length-to-diameter ratio is likely to be between 3:1 and 4:1, burner manufacturers[2,3] often providing data for guidance. Obviously, a very short furnace of large diameter, or a very long one of small diameter, is unsuitable.

Heat transfer in the furnace takes place mainly by radiation, where the heat flux (heat flow per unit area) is most intense in the boiler. The peak rate occurs at approximately one furnace diameter downstream from the burner front for oil or gas firing[4], rather more than this for coal.[5] It can reach a value of 320 kW m$^{-2}$ or even higher, so that high metal temperatures can prevail in this region. The peak heat flux depends on the cross-sectional area of the furnace, which is a function of the diameter, so that to avoid excess values, the permissible heat input to a furnace is related to the diameter. The relationship is shown in Fig. 12.1 which is reproduced from BS2790:1986, a similar diagram appearing in some European standards. It will be noted that for oil, gas and pulverised fuel firing the maximum heat input allowed per furnace is 12 MW, based on the net calorific value of the fuel, and for solid fuel fired on a grate, 8 MW. (See also section 4.1.3.) For inputs greater than these, two or more furnaces must be used. The mean heat flux in the furnace is generally a little over half the peak value, but is considerably more than that which occurs on other parts of the boiler except for the rear tube plate where, as will be seen later, local convection becomes important at the tube inlets.

It is clear that the metal of the furnace will be hotter than that of the shell and the tubular passes. It is also thicker (up to 22 mm), and is therefore very stiff. To avoid excessive stress on the end plates to which the furnace ends are attached, flexible elements must be provided in all but the smaller boilers. These take the form of corrugations, the whole or part

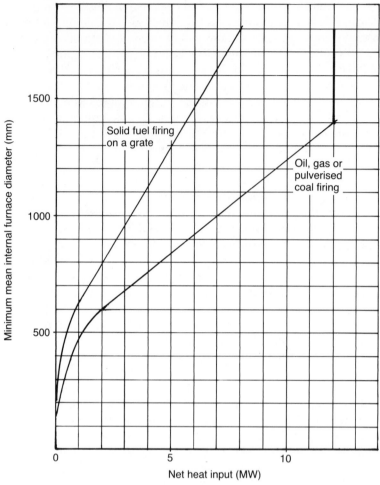

**12.1** Relationship between furnace diameter and permissible heat input
(From British Standard BS2790:1986, Fig. 3.1(6))

of the furnace being constructed in this manner or, more usually, separate
corrugations or '~       ~ hoops' are used at junctions of furnace sections.
The latter is s'         '2.2, and its design is fully explained in Section
3.10 of BS27                ion of the Standard also includes methods
for calculating t.              nace metal (which depends on the design
pressure), furnace ~             ethods by which it is supported, and
the specification of th.            onstruction. It also includes the
calculation of shell thick.             achment of tube plates, tubes
and standpipes for mountin~

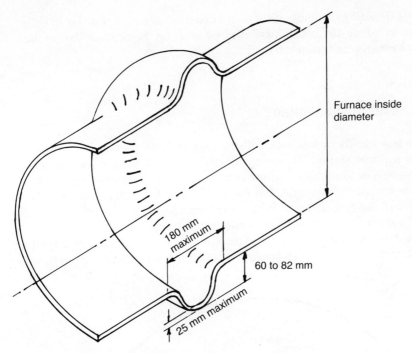

Furnace inside
diameter

180 mm
maximum

60 to 82 mm

25 mm maximum

**12.2** Furnace section with bowling hoop

Gas temperature measurements carried out at the furnace exits of boilers have resulted in the following empirical formula for its calculation.

$$T = k \, (H/A)^{0.25}$$ [12.1]

Where: $T$  is exit gas temperature (°C);
  $H$  is the heat input rate, based on the net calorific value of the fuel (W);
  $A$  is the furnace area exposed to radiant heat transfer (m²);
  $k$  is a fuel-dependent constant having values:

|                        |      |
|------------------------|------|
| for natural gas        | 52.4 |
| for gas oil            | 46.0 |
| for coal or heavy fuel oil | 38.0 |

The gas temperature will be high, varying from about 900 °C for coal to over 1200 °C for gas. Measurement errors can be serious at these high temperatures which are therefore based on suction pyrometer measurements. Moreover, the temperature can be very variable across the furnace exit, especially for coal firing on a grate.[5] It is therefore an approximation but is accurate enough for practical purposes. It will be noted that the fourth-power law of radiant heat transfer is involved, this mode of heat transfer being dominant in the furnace. (See also Ch. 5.)

Using the exit gas temperature thus calculated, the heat transfer in the furnace may be estimated, if needed, by deducting the heat content of the gases at furnace exit from the net heat input.

## 12.3   The reversal chamber

This is now mostly of wet-back construction and thus provides significant heat-absorption surface. Its dimensions must be such as to accommodate the first pass of tubes and to allow ample access for attention to the tube ends. In practice, this means that the area available for heat transfer is 30–40% of the furnace area. Where two furnaces are used, it is now usual to use a separate reversal chamber for each furnace, (see Fig. 12.3). In earlier days the practice was to use a common reversal chamber for both furnaces; this had the disadvantage of preventing the furnaces from being

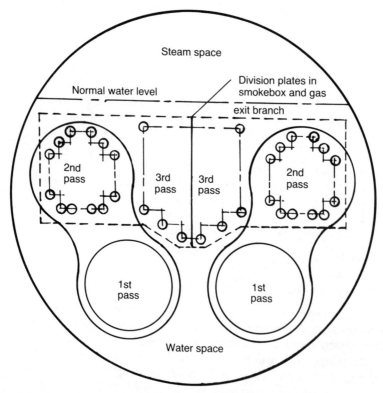

**12.3**   Twin-furnace firetube boiler with independent reversal chambers (From Wallsend Boilers Limited)

operated independently. The separate chambers permit the firing appliances to be adjusted to different firing rates, even to the extent of one being extinguished altogether. This allows a very wide turndown ratio to be accomplished.

Equation [12.1] can be used to determine the temperature of the gases entering the tubes, in which case the factor $A$ includes the effective areas of both the furnace and the reversal chamber. It is quoted in BS2790, but only the highest constant $k$, for gas, is allowed. This is because a boiler initially designed for coal or oil, may subsequently be used on gas, where the temperature of the gases entering the tubes is considerably higher than that for oil or coal firing. This may cause cracking and leaking at the tube ends unless due allowance is made at the design stage. This aspect is discussed more fully in the next section of this chapter.

Reversal chambers contain the lowest part of the heat-transfer surfaces in the boiler. When starting up from cold, condensation of water vapour will occur, not much with coal, quite copious amounts with oil, and very copious amounts with gas. The condensate will drain to the reversal chamber, so that a drain should be provided. With coal firing, much fly ash collects in the reversal chamber. This screens some of the heat-transfer surface, reduces $A$ (in equation [12.1]) and therefore increases tube entry gas temperature. This accumulation of ash can be a serious nuisance. Regular removal is desirable and adequate provision for this is not often made. Now that coal firing is increasing in importance, more consideration should be given to this problem; in earlier days when coal firing was very extensive, the bottoms of some dry-back boilers were constructed as a hopper from the bottom of which the fly ash could be drawn off by vacuum equipment. With wet-back boilers, a suitably sealed drop tube can be incorporated to allow dust to fall out and to pass from the reversal chamber for removal.

## 12.4   The tubes

The furnace and reversal chamber together comprise only about 10% of the heat-transfer surfaces of a boiler and yet account for some 40–50% of the heat transferred. The remaining surface and heat transfer occurs in the tubes, which can cool the gases to, say, 50 °C above the temperature of the water in the boiler. The design of the tubular passes, therefore, controls the efficiency of the boiler. By extending the tubular area, a lower exit temperature can be achieved. It takes much surface area to achieve this. It is more effective and more economical to install an economiser in which the water is at feed, not boiler, temperature. This provides a degree of countercurrent flow (see sections 5.6 and 7.4.1).

## 12.4.1 Heat transfer

Heat transfer by convection from gases flowing inside tubes has been dealt with classically by many authors and in Chapter 5 of this book. It has been shown by Roderick, Murray and Wall[6] that, when allowance is made for radiation, theory predicts practice with considerable accuracy. The same authors have also shown that calculations of draught loss through tube banks are realistic. The work described by these authors was carried out on sophisticated heat-transfer rigs and also on actual boilers, and the results are now widely used. Classical convection theory is based on dimensionless correlations for which the reader is referred to Chapter 5 and to the various textbooks on heat transfer. McAdams[7] deals with the subject in detail and suggests a dimensional formula. Converted into SI units and transposed, this becomes

$$t_2 = t_m + \cfrac{t_1 - t_m}{\text{antilog} \left\{ \cfrac{0.012\ 12L/D}{(DG)^{0.2}} \right\}} \qquad [12.2]$$

Where: $t_2$   is exit gas temperature (°C);
       $t_1$   is inlet gas temperature (°C);
       $t_m$   is metal temperature (water temperature plus, say, 20 °C) (°C);
       $L$   is tube length (m);
       $D$   is tube bore (m); and
       $G$   is mass flow rate per unit cross sectional area of tube bank available to gas flow ($kg\ m^{-2}\ s^{-1}$)

Whilst this formula provides an approximate answer, with boilers there is significant radiation from gases and some from solid particles. The procedure outlined later is recommended for boiler calculations. The above formula, however, clearly illustrates the importance of the length-to-diameter ratio of the tubes. As $L/D$ increases, so $t_2$, the temperature of the exit gas from the boiler, decreases and therefore efficiency increases.

In order to achieve an efficient boiler design, therefore, it is important that the tube bores should be as small as feasible. In practice, this means down to about 32 mm, below which tubes become difficult to fix in tube plates. Except in the smallest boilers it is generally better to use tubes of larger bore than this; fewer are needed and when fuels that may foul are used (coal and heavy fuel oil), larger tubes take longer to become blocked. However, to achieve a given efficiency, the tubes need to be longer. This is conveniently done in modern boilers by using two passes, otherwise longer boilers or smaller tubes will be needed, either of which may be inadmissible for the purpose for which the boiler is required. Against this, it must be said that in the past some very efficient boilers using only a single pass of relatively small bore (45 mm) tubes have been made, but they were expensive due to their length.

The work published by Roderick, Murray and Wall[6] presents the results in graphical form in Imperial units on logarithmic coordinates. The graphs are numerous, covering a wide range of tube diameters, and are widely used for the design of the tube passes of firetube boilers. The boiler exit temperatures predicted are very close to those measured in practice for all common fuels, so that this method of calculating heat transfer in boiler tubes is to be recommended. It is not possible here to reproduce all the graphs presented by those authors – the reader should refer to the original work.

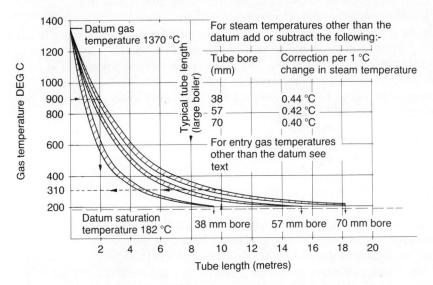

**12.4** Effect of tube length and diameter on exit gas temperature (Adapted From BCURA data document C/4813 (British Coal))

Figure 12.4 has been derived from three of the graphs for commonly used tube bores. This presents the temperature drop along tubes in terms of SI units on linear coordinates, which may be considered to present clearer pictures of the process than do logarithmic plots. Tubes of 38, 57, and 70 mm bore are considered for a total length of up to 18 m. It will be seen that there is a band for each tube size, the upper boundary of each band corresponding with a mass flow rate of gases in the tubes of 15 kg m$^{-2}$ s$^{-1}$, the lower boundary being 10 kg m$^{-2}$ s$^{-1}$. This represents a realistic range of operation at full load. The datum inlet gas temperature is 1370 °C and that for the steam and metal temperature is 182 °C, but the curves may be used for conditions other than these.

If a lower gas inlet temperature is used, as is usual, the appropriate curve is entered at that temperature, the corresponding tube length then being read. The total tube length is then added to this and the exit gas temperature for that length of tube is taken. For example, in Fig. 12.4, the top boundary curve for the 70 mm tube is entered at 900 °C and the corresponding tube length reading is 2 m. For a total tube length of 8 m, the exit gas temperature, 310 °C, is read at 2 + 8 = 10 m. The broken lines in Fig. 12.4 illustrate the procedure. The corrections for steam (metal) temperature are also shown on Fig. 12.4.

It will be seen that the effect of mass flow rate is small compared with those of tube length and diameter, which is consistent with equation [12.2.], where the exponent of $G$, the mass flow rate, is only 0.2.

It will be seen that the effects of tube length and diameter are strong. For the example taken, if 38 mm tubes were used instead of 70 mm, the exit gas temperature would be reduced from 310 °C to 210 °C, i.e. by 100 K, which is equivalent to a gain of about six efficiency points. The same reduction could be achieved for the 70 mm tube by increasing its length from 8 m to 17 m, needing a longer and more expensive boiler, indeed, impracticably so.

### 12.4.2   Draught loss

Modern firetube boilers present a considerable resistance to gas flow. In order to specify the fan pressures needed it is necessary to calculate this. Reference 6 provides a satisfactory method which uses the accepted laws of fluid flow, the essential component of which is that the pressure loss varies with the square of the mass flow rate (see equation [5.15]).

The furnace of a firetube boiler presents very little resistance to gas flow, little more than 1 mbar. The tubular passes on the other hand can result in a loss of 15 mbar or even more, which requires the consumption of significant electric power for driving the fans. It is desirable to minimise this, but to reduce it much below the 15 mbar mentioned would mean reducing the mass flow rate by increasing the cross-sectional area of the tubular passes. This would result in a very large and expensive boiler. In practice, it is a matter of compromise between boiler size and fan power (section 5.4).

Roderick, Murray and Wall present their procedure in the form of two graphs and a nomogram from which the entry loss, kinetic gain due to cooling in the tubes, and frictional resistance to flow along the tube can be read. These are given in Imperial units.

Figures 12.5, 12.6 and 12.7 express the results of Roderick, Murray and Wall in SI units. Together with Fig. 12.4, which is used to evaluate the gas temperature, they may be used to assess the draught loss through the tubes. Tube diameters may not fit exactly those used, but interpolation is valid for intermediate sizes. In addition to the losses given by these diagrams, there will be those due to change of direction of the gas flow in

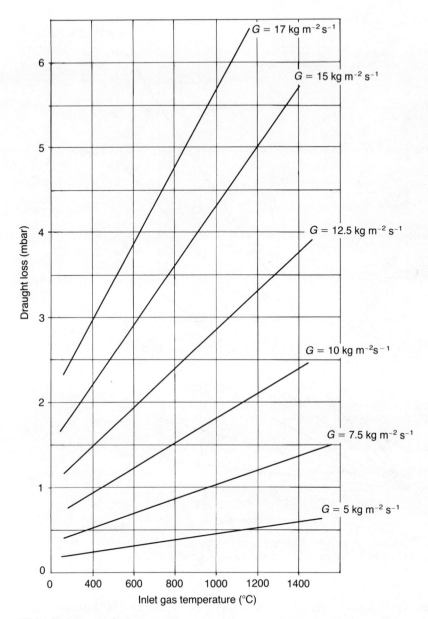

**12.5**  Tube entry loss
(After BCURA circular No. 192 (British Coal))

the reversal chamber. In the rear chamber the gas velocity will be
substantially that of the gases leaving the furnace. The mass flow rate is
quite low, around 2 kg m$^{-2}$ s$^{-1}$, and consequently the loss of pressure will
be low (less than 0.1 mbar), which may be neglected. This pressure loss is

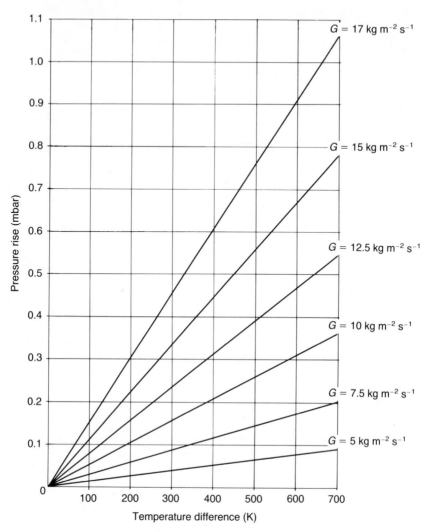

**12.6** Pressure rise in tubes due to cooling of the gas
(After BCURA circular No. 192 (British Coal))

based on the loss of two velocity heads ($V^2/2g$) due to two 90° changes. In
the front reversal chamber, however, the mass flow rate will be governed
by that in the tubes, which will be about 10 kg m$^{-2}$ s$^{-1}$. In this case the loss
due to the two velocity heads will be about 1.3 mbar. This is not great
compared with the tube losses and it would suffice to make an allowance of
1.5 mbar for three-pass boilers and none for two-pass boilers.

It is often beneficial, in boilers with three passes, to use tubes of
different bores in each pass: large bore tubes in the first pass to reduce

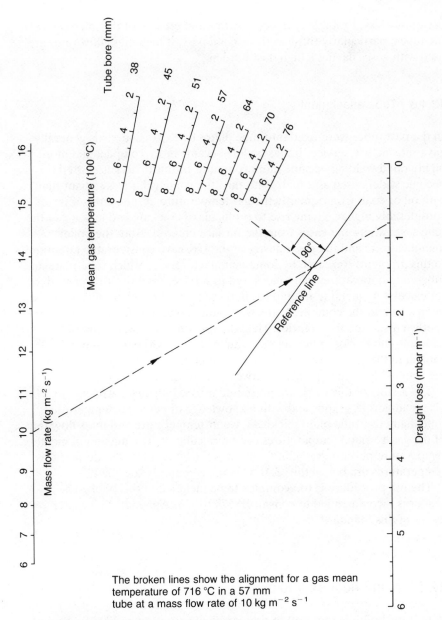

The broken lines show the alignment for a gas mean
temperature of 716 °C in a 57 mm
tube at a mass flow rate of 10 kg m$^{-2}$ s$^{-1}$

**12.7**  Nomogram for friction loss in tubes
(After BCURA circular No. 192 (British Coal))

fouling problems; smaller bore tubes in the second pass to ensure good
heat extraction and hence high efficiency in a boiler of reasonable length.
The same mass flow rate is used in each case. It is important that the mass
flow rate is not excessive, not much more than 10 kg m$^{-2}$ s$^{-1}$, and

preferably less. Problems can occur with rapid erosion of the inlet ends of the tubes, particularly those in the second pass, if high mass flow rates are used with gases having a high dust burden.

### 12.4.3 Tube attachment

In the past tubes were fixed into tube plates by expansion. This generally gave satisfactory service, but as boilers became more compact and more highly rated welding became more common. In the 1970s gas started to become widely used as a fuel. With this, the flame is far less radiant than with oil or coal. Consequently the gas temperature at tube entry was considerably higher, giving rise to problems of cracking and leakage at the tube seats where the gases from the furnace entered. Both the boiler manufacturers and the gas industry in the UK gave considerable attention to this problem. Research by some manufacturers, in which tube-plate and tube-end temperatures were measured in service, gave rise to a procedure for calculating metal temperatures in these parts. Local heat fluxes can be intense due to the combined effects of radiation and high convection coefficients at the tube entries. Heat dissipation to the water can be deficient due to close pitching of the tube seats, poor metal-to-metal contact in the seats, excessively thick tube plates and scale formation on the water side of the ligaments between the tubes.

The results of this work are presented in BS2790: 1986, Appendix C. Using numerous graphs and with a knowledge of entry gas temperature, tube diameter, tube plate thickness, water temperature and mass flow rate of the gases, metal temperatures can be calculated. The maximum metal temperature permitted is 420 °C, but it is recommended that design temperatures are below this, 380 °C being commonly specified.

The full procedure is too complex to be included in this book. The reader is recommended to consult BS2790:1986, Appendix C, or later issues of this Standard.

## 12.5   Numerical example

A firetube boiler is required to provide an evaporation of 6000 kg hr$^{-1}$ of steam at a pressure of 13 bar from feedwater at 80 °C. The efficiency (based on the GCV) at full load is to be 83%, using heavy fuel oil as fuel. The tube bore is to be 57 mm in both tubular passes. What furnace dimensions and tube lengths are required and what is the draught loss through the boiler?

## 12.5.1 Heat input needed

| | |
|---|---|
| Referring to steam tables[8], at 13 bar gauge pressure the latent heat of steam is | $1957.7 \text{ kJ kg}^{-1}$ |
| The saturation temperature is | $195 \text{ °C}$ |
| At this temperature the sensible heat is | $830.1 \text{ kJ kg}^{-1}$ |
| The feedwater temperature is | $80 \text{ °C}$ |
| At this temperature the sensible heat is | $334.9 \text{ kJ kg}^{-1}$ |
| The sensible heat to be added is $830.1 - 334.9 =$ | $495.2 \text{ kJ kg}^{-1}$ |
| The total heat requirement is $495.2 + 1957.7 =$ | $2452.9 \text{ kJ kg}^{-1}$ |
| Evaporating $6000 \text{ kg hr}^{-1}$ requires $6000 \times 2452.9 =$ | $14.717 \times 10^6 \text{ kJ hr}^{-1}$ |
| At the thermal efficiency of 83% the heat input in the fuel is | |
| $\dfrac{14.717 \times 10^6}{0.83} =$ | $17.73 \times 10^6 \text{ kJ hr}^{-1}$ |
| $=$ | $4.9255 \text{ MW (Gross CV)}$ |
| or $=$ | $4.65 \text{ MW (Net CV)}$ |

## 12.5.2 Furnace diameter needed

Referring to Fig. 12.1 (BS2790:1986, Fig. 3.1(6)) it will be seen that the net heat input is required for this purpose. The minimum furnace diameter required is 800 mm.

## 12.5.3 Furnace length required

It will be assumed that the furnace represents 0.75 of the total radiant heated surface, i.e. the effective area of the reversal chamber is 0.25 of the radiant heated surface. This assumption is conservative and may need modification as the design evolves and the boiler is drawn out. It will also be assumed that the tube entry gas temperature will be 950 °C which, from experience, represents good practice with heavy fuel oil firing.

Referring to equation [12.1] this states that

$$T = k(H/A)^{0.25}$$

i.e. $\quad 950 = 38 \left( \dfrac{4.65 \times 10^6}{A} \right)^{0.25}$

where $A = 11.9 \text{ m}^2$.

This is the total surface area, 0.75 of which is furnace area, i.e. $8.93 \text{ m}^2$.

For a furnace of 0.8 m diameter the length needed to provide this area is 3.56 m, giving a length-to-diameter ratio of 4.45 compared with a more usual figure gained from experience of, say, 3.2. The furnace diameter therefore needs to be increased to provide this. Thus

$$\pi D \times 3.2D = 8.93$$

whence

$$D = 0.942 \text{ m}$$
$$L = 3.2 \times 0.942 = 3.02 \text{ m}$$

It should be noted that this is also the length of the first pass of tubes. The boiler shell length will be governed by the length of the second pass of tubes needed to cool the gases to the required exit temperature to give an efficiency of 83%.

## 12.5.4 Exit gas temperature needed to give 83% efficiency

The boiler inefficiency, i.e. the sum of the individual losses, is made up of the following items (see section 11.2.2).

1. Dry gas loss — Unknown
2. Loss due to enthalpy in water vapour — Unknown
3. Loss due to incomplete combustion — 0.4% (legal stack solids limit)
4. Surface heat loss or radiation loss — 0.3% (BS845:1987, Part One, Table B3)

This leaves items 1 and 2 to total not more than 16.3%, to account for the heat input of 100%.

The formulae given in BS845 for these loss items with oil firing are:

1. $L_1 = \dfrac{k}{V_{CO_2}}(t_3 - t_a)$  [12.3]

2. $L_2 = \dfrac{(m_{H_2O} + 9H)\,(2488 - 4.2t_a + 2.1t_3)}{Q}$  [12.4]

Where: $K$  is the Siegert constant (for oil this is 0.51);

$V_{CO_2}$  is the percentage by volume of $CO_2$ in the dry flue gases, assume 13%;

$t_3$  is the boiler exit gas temperature, assume saturation temperature + 50 °C = 195 °C + 50 °C = 245 °C, say 250 °C in the first instance;

$t_a$  is the temperature of air entering the boiler, used as the datum for the calculation, (assume 15 °C);

$m_{H_2O}$  is the percentage water in the fuel;

$H$  is the hydrogen content of the fuel (%); and

$Q$  is the gross calorific value of the fuel.

Where a fuel analysis is given by the purchaser, $m_{H_2O}$, $H$ and $Q$ will be obtained from this. In this instance, the typical analysis given in Table 3.3 will be used where the values are 0.1%, 11.4% and 42 900 kJ kg$^{-1}$, respectively. Substituting the values in equations [12.3] and [12.4]

$$L_1 = \frac{0.51}{13}(250 - 15) = \qquad\qquad 9.22\%$$

$$L_2 = \frac{(0.1 + 9 \times 11.4)(2488 - (4.2 \times 15) + (2.1 \times 250))}{42\ 900} = \qquad 7.06\%$$

Total                                                                    16.28%

The exit gas temperature is therefore approximately correct. In practice, the temperature first assumed may give a significantly different efficiency, and this necessitates further calculations until a sufficiently close approximation is achieved.

For regularly used, consistent fuels, such as oil, natural gas and selected coals, values of efficiency for various exit gas temperatures will be obtainable from curves produced by design engineers from previous calculations. This avoids the laborious repetition of routine calculations. Figure 13.1 for natural gas is a typical example.

### 12.5.5   The tube length needed

It will be assumed that sufficient tubes will be used in each pass so that the mass flow rate of the gases is 10 kg m$^{-2}$ s$^{-1}$. It was shown earlier that the gross heat input is $17.73 \times 10^6$ kg hr$^{-1}$ so that the mass input of fuel oil is

$$\frac{17.73 \times 10^6}{42\ 900} = 413.3 \text{ kg hr}^{-1}$$

(42 900 is the gross calorific value of the fuel oil in kJ kg$^{-1}$).
The mass of the products of combustion per mass of fuel may be calculated by the method shown in Chapter 3, or be taken from previous calculations or from published information.[9] Often, the stoichiometric air requirement for a fuel is expressed in terms of mass of air per unit of calorific value.[10] This enables the requirement to be readily adjusted for small variations of calorific value from that for which the values have been calculated.

Using published data for the fuel oil analysis used, it is found that the stoichiometric air required is 13.84 kg per kg of fuel.[11] Using Fig. 3.1 it is found that 13% $CO_2$ is equivalent to 20% excess air. The mass of the products of combustion per mass of fuel therefore becomes:
Mass of fuel + 1.2 × (mass of stoichiometric air)

$$= 1 + 1.2 \times 13.84$$
$$= 17.61 \text{ kg per kg oil.}$$

The gas flow through the boiler will therefore be

$$\frac{17.61 \times 413.3}{3600} = 2.02 \text{ kg s}^{-1}$$

If the mass flow rate through the boiler tubes is 10 kg m$^{-2}$ s$^{-1}$, the total area to gas flow will be

$$\frac{2.02}{10} = 0.202 \text{ m}^2.$$

If the tube bore is 57 mm (0.057 m) the number of tubes will be

$$\frac{0.202}{\dfrac{\pi}{4} \ (0.057)^2} = 80.$$

Reference may now be made to Fig. 12.4 (or the original work in f.p.s. units), entering the 57 mm bore tube curves at 950 °C, equivalent to 1.3 m tube length, the required exit gas temperature is 250 °C. Reading the curves to the right it will be seen that the total tube length needed for a 57 mm tube is 10.0 m. The required tube length is therefore 10 − 1.3 = 8.7 m. The furnace length is 3 m, which corresponds with the length of the first-pass tubes. This means that the second-pass tubes need to be 5.7 m long if the intended exit gas temperature is to be achieved. A length of about 1 m is required for the reversal chamber; a second-pass tube length of 4 m would therefore be more appropriate to optimise the boiler.

Reference to the 38 mm curve shows that 250 °C can be obtained with only 5.6 m total tube length, which would make for a compact boiler. The bore is a little small, however, for modern heavy fuel oils, some of which may tend to foul. There could be a case, therefore, for using 57 mm tubes in the first pass and 38 mm in the second pass since fouling takes place mainly at the inlets of the first pass of tubes. The estimation of exit gas temperature needs to be done in two stages.

1. First pass, 57 mm tubes, 3 m long
   Inlet temperature                                              950 °C
   Outlet temperature                                            480 °C
2. Second pass, 38 mm tubes;
   Inlet temperature                                            480 °C
   Length needed for 250 °C outlet temperature

                                                  3.8 (i.e 6.5 m − 2.7 m)

This leaves 0.8 m for the accommodation of the reversal chamber, which is insufficient. The effective second-pass tube length selected will therefore be 4 m, giving an exit gas temperature of rather less than 250 °C and a small gain in efficiency, which is to the good. This arrangement will therefore be adopted.

Having obtained the tube lengths, their diameters and the inlet and outlet temperatures, the draught loss may now be calculated using Fig. 12.5, 12.6 and 12.7 as follows.

For the first-pass tubes, 57 mm bore, 3 m long, use Fig. 12.5 (entry loss). Enter at 950 °C and read to the 10 kg m$^{-2}$ s$^{-1}$ line. This gives          1.75 mbar

Using Fig. 12.6 (pressure rise due to cooling), the temperature difference is $950 - 480 = 470$ K. Reading to the 10 kg m$^{-2}$ s$^{-1}$ line gives          0.24 mbar

The net pressure loss is                                                  1.51 mbar

Using Fig. 12.7, mean temperature is          $\dfrac{950 + 480}{2}$          $= 715$ K

Drop a perpendicular from 715 °C on the 57 mm tube line to the reference line, as shown by the broken lines on Fig. 12.7. Align the mass flow rate of 10 kg m$^{-2}$ s$^{-1}$ with point of the intersection on the reference line and project it to read the draught loss per metre of tube.          0.8 mbar m$^{-1}$

For a tube 3 m long, draught loss is                                      2.40 mbar

The total draught loss on the first-pass tube $= 1.51 + 2.40$          $= 3.91$ mbar

For the second-pass tube, 38 mm bore, 4 m long, proceed as for first-pass tubes. Using Fig. 12.5, enter at 480 °C and read to the 10 kg m$^{-2}$ s$^{-1}$ entry loss                                                                  1.05 mbar

Using Fig. 12.6, the temperature difference is

$480 - 250 =$                                                            230 K

The pressure rise is                                                      0.12 mbar

The net pressure loss is                                                  0.93 mbar

Using Fig. 12.7, mean temperature is          $\dfrac{480 + 250}{2}$          $= 365$ K.

Enter the 38 mm tube line at 365 K and, as before, read          0.8 mbar m$^{-1}$

For a 4 m tube length draught loss is                                     3.20 mbar

The total draught loss on the second-pass tubes $= 0.93 + 3.20 = 4.13$ mbar

The total draught loss over the tubes is $4.13 + 3.91 + 0.15$ (for reversals)
$= 8.19$ mbar (3.29 in WG)

This provides an efficient boiler with fairly low draught loss, but the number of tubes in the second pass will be greater than that in the first pass. This number can be trimmed by increasing the mass flow rate of the gases in the second tubular pass at the expense of a small rise in exit gas temperature and draught loss. This will provide a boiler of somewhat smaller diameter, which would be cheaper. It will now be necessary to draw out the boiler using tube spacings as required by BS2790:1986, in order to obtain the shell dimensions. A typical diameter for this size of boiler would be about 2.5 m, with an overall length of 5.7 m including burners and gas outlet duct.

## 12.6 Steam release velocity

The shell diameter is about 2.5 m and its length, determined by the length of the second pass of tubes, is 4.0 m. A reasonable figure for the water level is about 20% of the diameter below the crown. The final figure will need to be determined after the boiler has been drawn out, since all the heated surfaces, and particularly the crown of the rear reversal chamber, will need to be well covered by water (by at least 100 mm). This is to give a margin to avoid overheating should a low water level accidentally occur.

From Fig. 12.8 it can be determined that the width of the steam release surface at 20% of the diameter below the crown is 80% of the diameter, which is 2.0 m. With a boiler length of 4 m, the water plane area is 8 m².

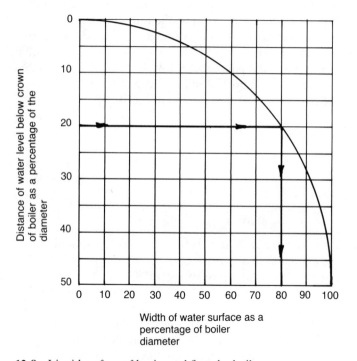

**12.8** Liquid surface of horizontal firetube boilers

From section 12.5, the design evaporation is 6000 kg hr⁻¹ at 13 bar pressure. From steam tables the specific volume is 0.141 m³ kg⁻¹, so that the volumetric evaporation is

$$6000 \times 0.141 \quad = 846 \text{ m}^3 \text{ hr}^{-1}$$
$$= 0.235 \text{ m}^3 \text{ s}^{-1}.$$

The rate of steam release from the water plane is

$$\frac{0.235}{8} = 0.0294 \text{ m s}^{-1}$$

This is well within the limit set down in section 12.1 ($0.056 \text{ m s}^{-1}$), and is therefore acceptable. Had the pressure been lower, say 7 bar, the volume of steam generated would have been 1764 $\text{m}^3 \text{ hr}^{-1}$ and the release rate would be $0.061 \text{ m s}^{-1}$; this would require a rather longer shell.

## 12.7  Rear tube plate temperature

This was discussed in section 12.4.3. There is no alternative to carrying out the full calculation procedure given in Appendix C of BS2790:1986. This cannot be done until the boiler has been drawn out and the tube plate thickness established, which is a vital factor in the calculation. It is often found that, in an initial design, the tube plate thickness, is such as to take the metal temperature beyond the recommended limit of 380 °C. This has to be corrected by decreasing the thickness and using more stay tubes. This subject must be studied in detail in BS2790:1986.

## 12.8  Summary

In this chapter, the design of a firetube boiler for a given output and efficiency has been considered. The following leading data have evolved.

| | |
|---|---|
| Output | 6000 kg $\text{hr}^{-1}$ at 13 bar pressure |
| Efficiency | 83% on the gross calorific value |
| Furnace diameter | 0.943 m |
| Furnace length | 3.02 m |
| First-pass tube bore | 57 mm (62 in number) |
| Second-pass tube bore | 38 mm (178 in number, but fewer are permissible at the expense of some increased draught loss) |
| Draught loss over boiler | 8.19 mbar |
| Temperature of gas entering first-pass tubes | 950 °C |
| Temperature of gas entering second-pass tubes | 480 °C |
| Exit gas temperature | 250 °C |

# References

1. Department of the Environment. Property Services Agency. Specification M and E3.
2. Trade literature. H. Saacke and Co. Ltd, Farlington, Portsmouth.
3. Trade literature. Hamworthy Engineering Ltd, Poole, Dorset.
4. Gunn, D. C. 'Heat transfer in the furnaces of shell type boilers'. *Proc. Inst. Mech. Eng.* 187 64/78, p. 809.
5. Wright, S. J. 'Temperature and heat transfer patterns in chain-grate stoker-fired economic boilers'. *J. Inst. Fuel* 1965 (Mar.) p. 124.
6. Roderick, D. J. I., Murray, M. V. and Wall, A. C. 'Heat transfer and draught loss in the tube banks of shell boilers', *J. Inst. Fuel* 1959 (Oct.) **32**: p. 450.
7. McAdams, W. H. *Heat transmission* (3rd edn). McGraw-Hill. p. 227.
8. Steam tables evaluated from the 1967 International Formulation Committee Formulation for Industrial Use. NEI International Combustion.
9. Rose, J. W. and Cooper, J. R. *Technical Data on Fuel* (7th edn), 1977. British National Committee, World Energy Conference, London.
10. Johnson, A. J. and Auth, G. H. *Fuels and Combustion Handbook*, McGraw-Hill. p. 355.
11. Rose, J. W. and Cooper, J. R. *Technical Data on Fuel* (7th edn), 1977. British National Committee, World Energy Conference, London.

# CHAPTER 13

# Water-tube boilers – design for output, efficiency and reliability

## 13.0 Introduction

As mentioned in Chapter 12, most boilermakers have established standard designs of boiler from which the unit for a particular application is selected. For the commonly used fuels there also tend to be established operating criteria such as excess air for combustion and final flue gas temperature. The design engineer will have available to him the results of previous studies and designs, it therefore becomes unnecessary to carry out all the processes detailed in this chapter for every project.

## 13.1 Output and operating conditions

These will normally be specified by the user, or his representative if using a consultant or other engineering service, after due consideration of all the aspects detailed in Chapter 2.

There is a tendency in the water-tube boiler industry, particularly for the larger units, to refer to the output of boiler/turbo generator combinations in terms of the electrical output of the turbo generator in megawatts. For turbines generating power only using condensing turbines of reasonable efficiency, this can be about 35% of the heat input to the steam in the boiler. It is currently more usual therefore with water-tube steam generators to give the boiler output or evaporation in kg s$^{-1}$ or t hr$^{-1}$. To establish the heat input to the steam it is necessary to know the operating pressure and steam temperature required at the boiler outlet and the temperature of the feedwater provided at the boiler inlet or at the supplier's terminal point. Given these conditions, details of the range of fuels to be fired and as much of the information discussed in Chapter 2 as possible, the boilermaker will be able to determine the type of combustion equipment and most suitable boiler to use.

## 13.2   Efficiency and fuel consumption

Before selecting the dimensions of a boiler and firing equipment it is necessary to determine the thermal efficiency and fuel consumption. On occasions the thermal efficiency or flue gas temperature to be used are specified by the user, but frequently it is left to the boilermaker to determine the values that the boiler is to achieve. The thermal efficiency of a boiler is the percentage of the heat supplied by the fuel that is transferred to the steam by the boiler (Ch. 11). For design purposes it is calculated by the indirect or losses method, that is the various heat losses are determined as a percentage of the heat supplied by the fuel, the total is then subtracted from 100 to give the percentage of the heat transferred to the steam.

The heat losses from a boiler are listed in Chapter 11, as are the methods of calculating the thermal efficiency from the results of detailed tests carried out on a boiler after commissioning. These calculations are much simplified at the design stage, the design engineer being able to select the various criteria of operation determined largely from experience.

### 13.2.1   Determination of thermal efficiency

#### 13.2.1.1   Heat loss to flue gases

The heat lost to the flue gases depends upon the quantity of products of combustion and the temperature of the gas leaving the heat-recovery equipment. The products of combustion for the fuel to be used are determined by the principles described in Chapter 3. Such routine calculations are carried out using computer facilities where these are available or alternatively by graphical means for the commonly used fuels.

The quantity of excess air to be employed to ensure complete and satisfactory combustion is determined from knowledge of the combustion characteristics of the fuel and the type of combustion equipment to be used; previous experience plays an important role in this. Typical values for industrial watertube boilers where variable loads are experienced are:

| | |
|---|---|
| natural gas | 10% |
| heavy fuel oil | 10% |
| coal, stoker fired | 35 to 40% |

Lower values can be used, but they require good maintenance and operation of the firing equipment for them to be sustained.

When selecting the flue gas temperature that is to be achieved, the following factors must be considered.

Boiler availability, i.e. the percentage of the annual working hours for which the boiler is available for use:

fuel costs
type and composition of fuel
feedwater temperature available

The effects of these are interrelated to some extent. For instance, a boiler giving high thermal efficiency does not necessarily give a high availability if a low flue gas temperature is used which results in rapid fouling of the heat-recovery equipment.

Low-temperature corrosion, explained in section 10.4.2, has a significant influence upon the choice of final gas temperature when heat-recovery equipment is included. A combination of gas outlet and cold fluid inlet temperatures must be selected to give the desired minimum tube surface temperature within the heat-recovery equipment.

If low-temperature corrosion is to be avoided and a long life is desired for the heat-recovery equipment, it is recommended that the flue gas temperature should not be less than 180 °C for sulphur bearing oils and 160 °C for sulphur bearing coals. Where specific fuel analyses are available, dewpoint temperatures can be determined, (equation [10.6]), and a more applicable flue gas temperature selected to suit the prevailing conditions. When using economisers it is rarely economical to reduce the flue gas temperature to a value less than 30 °C above the inlet temperature of the feedwater.

The values of the flue gas heat losses are estimated per unit mass of fuel as a percentage of either the gross or net calorific value, as explained in section 3.5, and are calculated above a datum temperature (usually the prevailing ambient air temperature). For the purposes of this chapter the method using the gross calorific value will be dealt with.

The expressions used are as follows.

heat loss to dry flue gas

$$= \frac{W_g}{Q} \times C \, (t_g - t_a) \qquad\qquad [13.1]$$

Heat loss to moisture in flue gas formed by combustion of the Hydrogen and evaporation of the moisture in the fuel

$$= \frac{W_M}{Q} \times (Ht_g - ht_a) \qquad\qquad [13.2]$$

Heat loss to moisture in flue gas from humidity in combustion air and from moisture in fuel where this exists as a vapour in the fuel

$$= \frac{W_m}{Q} \times (Ht_g - Ht_a) \qquad\qquad [13.3]$$

Where: $W_g$   is the mass of dry flue gas per unit mass of fuel,
        $W_M$ is the mass of moisture from hydrogen and water in the fuel per unit mass of fuel;
        $W_m$ is the mass of moisture from water vapour in the combustion air and fuel per unit mass of fuel;
        $C$   is the mean specific heat of the dry flue gas at the flue gas temperature;
        $t_g$   is the flue gas temperature;
        $t_a$   is the ambient temperature;
        $Ht_g$ is the enthalpy of water vapour at flue gas temperature;
        $Ht_a$ is the enthalpy of water vapour at ambient temperature;
        $ht_a$ is the enthalpy of water at ambient temperature; and
        $Q$   is the gross calorific value of the fuel.
(All in self-consistent units.)

### 13.2.1.2   Heat lost to unburnt combustibles

Allowance is rarely made in the design for losses due to incomplete combustion when firing natural gas and fuel oils unless the latter have a high ash content. With solid fuels unburnt carbon is contained in the residuals discharged from ash and grit hoppers and the chimney. The value depends upon the combustion system used and is determined from experience with the operation of boilers under similar conditions. The value selected also makes allowance for the sensible heat carried away in the ash and grits, where necessary.

### 13.2.1.3   Heat lost to the atmosphere from the boiler walls (radiation loss)

While BS2885[1] indicates methods of determining this, any theoretical calculations can only be approximate due to the many factors involved in determining the surface area of the boiler and equipment from which heat is lost and the rate at which it is lost from that surface. For a given output a coal-fired boiler will be larger than one fired with gas or oil, and hence the radiation loss will tend to be higher with coal firing (Table 13.1).

Reasonable values of heat lost have been determined over years of operation, and the value to be used in efficiency tests is eventually agreed between the supplier and purchaser. BS845:1987[2] tabulates typical values for water-tube and firetube boilers, but makes no differentiation for the fuel being fired. An easy means of estimating a value is also given in the ASME test code.[3]

Because the temperature of the outer surface of a boiler enclosure does not reduce significantly as the steam demand upon it reduces, the heat lost from the surface will be substantially the same over the whole load range of the boiler. As a percentage of the heat input to the boiler, the radiation loss will therefore increase as the load on the boiler reduces and is usually taken as inversely proportional to load. For example, the percentage heat

**13.1**   Typical values of thermal efficiency

|  |  | Natural gas | Oil | Coal |
|---|---|---|---|---|
| Flue gas temperature | (°C) | 125 | 180 | 160 |
| Excess air supplied | (%) | 10 | 10 | 35 |
| Dry gas loss | (%) | 3.39 | 5.68 | 6.21 |
| Moisture loss | (%) | 10.67 | 6.56 | 4.90 |
| Unburnt combustible loss | (%) | 0.00 | 0.10 | 2.00 |
| Moisture in air loss | (%) | 0.07 | 0.11 | 0.12 |
| Radiation loss | (%) | 0.60 | 0.60 | 1.20 |
| Total Losses | (%) | 14.73 | 13.05 | 14.43 |
| Efficiency based on Gross Calorific Value | (%) | 85.27 | 86.95 | 85.57 |

loss by radiation at 50% boiler output will be approximately twice that at 100% load.[3]

For a given fuel and method of firing the heat lost to radiation as a percentage of the heat supplied in the fuel reduces as the design output, and hence physical size of a boiler, increases.[2,3] This gives a marginally higher efficiency at high design outputs, all other factors being equal. This is explained by the fact that the output capability of a boiler is approximately proportional to its volume, but the surface area of the boiler increases at a lower percentage rate does than the volume, and hence the quantity of heat lost increases at a lower rate than does the output. Typical variations of radiation loss with boiler size and output are given in Table 13.2.

**13.2**   Variation of radiation loss with boiler size and at part loads as a percentage of the heat input

| 100% boiler output (kg s$^{-1}$) | 7 | 14 | 21 |
|---|---|---|---|
| Radiation loss at 100% output | 0.72 | 0.54 | 0.48 |
| Radiation loss at 75% output | 0.96 | 0.8 | 0.57 |
| Radiation loss at 50% output | 1.32 | 0.96 | 0.9 |

### 13.2.1.4   Factors affecting thermal efficiency

Typical values for the heat losses and thermal efficiency are given in Table 13.1 for natural gas, oil and coal having analyses as given in Chapter 3, and Tables 3.1 and 3.3. As will be seen from Table 13.1, by far the greatest heat loss is that carried away by the flue gases, a reduction of the flue gas temperature of 20 °C is equivalent to an increase in thermal efficiency of approximately 1%, e.g. an increase of efficiency from 85% to 86%.

The quantity of flue gases should be maintained at the minimum possible by good combustion control and elimination of unwanted air infiltration through the boiler enclosure by good maintenance.

When a boiler is fitted with economisers or airheaters which are designed to give the selected flue gas temperature, the boiler working pressure has no influence upon the thermal efficiency. When no heat recovery is included, for a given heating surface and heat input to steam, a boiler operating at a high pressure will give a higher temperature of gas leaving the convection heating surfaces than if it is operating at low pressure. This is due to the higher saturation temperature of the water in the boiler tubes giving a lower LMTD and hence lower heat transfer within the convection surfaces (section 5.6).

The heat loss to unburnt combustibles in the ash and grit may be high due to either a high ash content or a very high carryover of particulates of high carbon content into the boiler, as experienced with spreader stokers and fluidised beds (Ch. 14). Some reduction of this loss, and hence improvement of efficiency, can be obtained by refiring or reinjection of the grits from the boiler hoppers into the furnace to burn off some of the remaining carbon. Two-stage dust collectors (in series on the gas side) are often included in such cases, and the coarse material from the first stage only is refired.

Complete refiring of all the grits should not be carried out as this will increase the quantity of grit flowing across the boiler heating surfaces significantly, hence increasing the probability of erosion of the tubes. Also, with refiring the grits are reduced in size with the consequence that a more efficient dust collector is required or the chimney particulate emission will increase.

In the event of the required boiler efficiency being specified in an enquiry to the boilermaker, the design engineer has to determine the flue gas temperature required to give this efficiency before the amount of heating surface in the heat recovery equipment can be determined. This calculation will be by trial and error as for firetube boilers (Ch. 12), although for consistent fuels such as natural gas and oil, results can be obtained rapidly from curves produced from previous calculations (see Fig. 13.1).

As the output from a boiler reduces below the design of 100% the values of the losses vary and will result in a change of boiler efficiency. The effects upon the losses of reducing boiler output are as follows.

1.  Unless action is taken to prevent it for corrosion reasons, the flue gas temperature will fall, giving a reduction of the dry flue gas and moisture losses.
2.  The excess air in the combustion chamber, and hence oxygen content

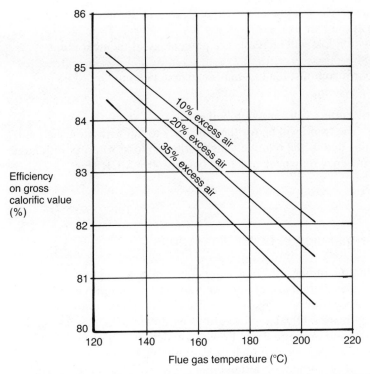

**13.1** Variation of efficiency with flue gas temperature and excess air with natural gas firing

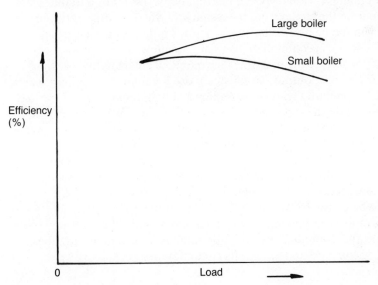

**13.2** Variation of efficiency with load

in the flue gases, will increase (Fig. 3.1), thus increasing the dry gas and moisture in air losses.

3. The radiation loss will increase as a percentage of the heat input (13.2.1.3).

4. The unburnt combustible loss may change depending upon the fuel and firing system.

The cumulative effect will vary with boiler size, fuel and final gas temperature. The trend will be for the efficiency to fall with reducing load, with small units due to the relatively high radiation loss, whereas with large boilers the efficiency may peak at a load below 100% MCR (Fig. 13.2).

## 13.3   Determination of boiler dimensions

When selecting the dimensions of a water-tube boiler for a given application, the availability of pre-engineered designs and previous experience is of considerable assistance. The height and the width across the gas flow path are interrelated by various parameters such as gas and steam velocities and pressure drop. Previous experience enables one of these dimensions to be selected such that acceptable results will be obtained with the minimum of design input.

At this stage it is necessary to consider the dimensions of the steam drum required because this can also influence the dimensions of the boiler. With water-tube boilers, various types of internal baffle are utilised in the steam drum to separate the steam from the steam and water mixture returning to the drum from the heated boiler tubes. These baffles require a certain amount of space to accommodate them and to allow access for inspection and fitting. Because of the high specific volume of steam at low pressure (Fig. 6.2), for a given output, larger steam drum volumes are required at low pressures compared with those required at high pressures.

Depending upon the type and arrangement of separator used there is some flexibility in the selection of the steam drum dimensions, for example for a given volume the drum can vary from a short one of large diameter to a long one of small diameter. There is a minimum diameter of about 1060 mm into which any type of internal separator can be accommodated and also allow access for its installation. The drum size required has therefore to be considered early in the design of the boiler, particularly when using multi-drum boilers having a bank of boiler tubes between the drums where the drum diameter will also influence the number of tubes that can be accommodated around its circumference (Fig. 13.3).

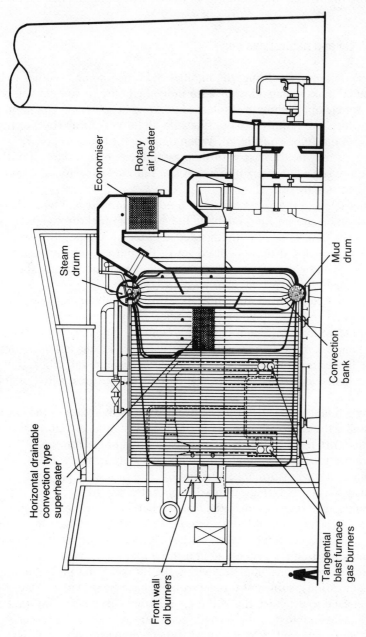

Economiser

Rotary
air heater

Steam
drum

Mud
drum

Convection
bank

Horizontal drainable
convection type
superheater

Front wall
oil burners

Tangential
blast furnace
gas burners

**13.3**   Two-drum oil-/gas-fired boiler

## 13.4 Furnace dimensions

These are influenced to a great extent by the type of firing equipment used.

### 13.4.1 Oil and natural gas firing

For oil and natural gas firing, the number of burners to be used has to be determined making due allowance for individual burners to be taken out of service for cleaning while the boiler is operating at full load. The minimum load at which it is required to operate with all burners firing (burner turn-down) also has to be considered. When only one or two burners are installed, as may be the case with shop-assembled boilers, the first of these may require special burners into which two full-load capacity oil guns can be inserted side by side.

Having selected the excess air to be used, the number and type of burners can be chosen, thus enabling the maximum fuel throughput per burner to be determined. With knowledge of the air temperature available and the air-pressure drop to be used across the burner, the dimensions of the flame given by the selected burner can be determined from data obtained from previous boilers or test rigs.

The burner centre-to-centre and centre-to-side-wall dimensions can be established. These, combined with the selected burner configuration (i.e. the number wide and the number high), enable the dimensions of the flame envelope and furnace to be fixed. The furnace size is selected to ensure that the flames do not impinge upon the furnace water-wall tubes, thus causing excessive local heat fluxes and metal temperatures, with possible tube failure.

Firing a boiler with the burner centre lines at right angles to the gas flow, as would be the case with burners in one side wall of a conventional boiler (illustrated in Fig. 5.4), can cause maldistribution of gas flow and temperature gradients through the boiler, particularly at low loads. This can reflect on the performance and metal temperatures of pendant superheaters, necessitating careful design of these.

When using corner or tangential firing, where a vertical row of burners is installed in each corner of the furnace (see Fig. 13.4), the heat input per unit of plan area becomes a factor in the selection of furnace dimensions. To give the desired residence time, ensuring that satisfactory combustion has taken place and is completed within the furnace (thus giving minimum particle emission), sufficient furnace volume must be provided. One criterion used to establish this is the volumetric heat release rate, that is the total heat supplied to the furnace per hour divided by the furnace volume and expressed as $W\ m^{-3}\ hr^{-1}$. When comparing boiler designs care must be taken to ensure that values quoted are all on the same basis for

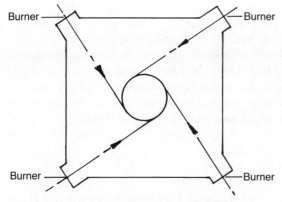

**13.4**   Principle of tangential (corner) firing

calculation of the furnace heat input, i.e. the fuel consumption multiplied by either the gross or net calorific value of the fuel. Typical values of heat release are given in Table 13.3.

**13.3**   Typical values of volumetric heat-release rate

| Oil firing | Shop-assembled | $0.6 - 0.8$ MW m$^{-3}$ |
|---|---|---|
| | Site-assembled | $0.3 - 0.4$ MW m$^{-3}$ |
| Natural gas | | $0.3 - 0.46$ MW m$^{-3}$ |
| Coal | | $0.16 - 0.26$ MW m$^{-3}$ |
| Blast-furnace gas | | $0.25$      MW m$^{-3}$ |

With oil and gas firing, particularly in highly rated and large boilers, the maximum heat flux to the furnace wall tubes must be considered to ensure that no boiler water circulating problems or overheating of the wall tubes will occur in the furnace (section 5.11).

### 13.4.2   Solid fuel firing – mechanical stokers

On boilers firing solid fuels using mechanical stokers, the dimensions of the stoker required to satisfactorily burn the fuel will establish the furnace plan dimensions. There may be dimensional modules of stoker width and length that it is required to maintain to utilise pre-engineered designs. If so, these have to be considered. For travelling-grate stokers, the width of the grate must be such that the speed of travel of the fuel through the furnace does not exceed its speed of ignition, otherwise the point at which ignition commences will eventually travel along the stoker to the discharge end and the fire will be extinguished.

Conservative grate and furnace ratings give flexibility with the range of fuels that can be used. In the long term there is a tendency for the quality of coals used on industrial water-tube boilers to deteriorate as the sources of better quality coals are exhausted. Care must be taken to ensure that, when alternative sources of coal are considered, the combustion equipment is capable of firing them and that they do not have constituents or properties for which the stoker and boiler are unsuitable. It is very easy when considering fuel costs to be tempted by the low cost of a grade of coal which may result in a loss of boiler output and availability.

### 13.4.3   Pulverised coal firing

Selecting the furnace dimensions for boilers fired with pulverised coal is carried out in a manner similar to gas and oil firing where the number of burners and flame dimensions have to be considered. Furnace plan area rating again plays a part in the selection. With any method of solid fuel firing the ash fusion and fouling characteristics must be taken into consideration. Where practicable, the temperature of the gases leaving the furnace should be below the initial deformation temperature of the ash, i.e. the temperature at which the ash in the fuel starts to soften and hence become 'sticky'. This is to prevent buildup of the ash on the relatively closely spaced tubes of the superheater and evaporative heating surfaces at the furnace outlet (see Ch. 10).

With low ash fusion fuels, or with boilers giving high degrees of superheat, it may be necessary for the gas temperature to be at or above the initial deformation temperature of the ash to be able to maintain a practical temperature difference within the superheater. In such cases, wide tube spacings in the superheater at right angles to the gas flow will be necessary to avoid bridging and blockage of the gas passages with molten ash, and so to maintain good boiler availability. Past experience is essential in such cases to ensure that the correct tube spacings are selected.

## 13.5   Superheaters

It is necessary to maintain a reasonable temperature difference between the flue gases and steam in the superheater to give an acceptable heating surface and hence superheater dimensions. Superheaters are therefore usually installed before the major mass of evaporative convection surfaces (Fig. 5.8). Exceptions are boilers having more than one convection bank (Fig. 1.10), where only small degrees of superheat are required to give dry steam at the consumer, when the superheater may be placed at the boiler outlet. The combination of high metal and gas temperatures makes the superheater susceptible to fouling (section 10.1 and 10.2) and possibly to high-temperature corrosion in large high-pressure boilers where high steam temperatures are being used.

Superheaters can be of the radiant type, in which they are exposed to furnace radiation and the majority of the heat is transferred by radiation (e.g. the secondary section of Fig. 2.5). Alternatively, they can be of the convective type, where they are shielded from furnace radiation as illustrated by Fig. 13.3, or they can be a combination of the two, as illustrated by Fig. 2.5. The convective type of superheater is favoured for the rigours of industrial applications, operating metal temperatures generally being less than for the radiant type for the same steam conditions.

There is also the choice of vertical-tube pendant superheaters with horizontal gas flow (Fig. 13.5), or horizontal-tube superheaters with vertical downward gas flow (Fig. 13.6). Horizontal tubes with vertical upward gas flow are possible in suitably designed boilers. The vertical-tube or horizontal-tube with vertical upward gas flow types are most suitable under gas-side fouling conditions, because any deposits that may settle out from the gases, or those that having adhered to the tubes, break away and fall off, will fall clear of the tubes. With the horizontal type having vertically downward gas flow, any deposits which form on the gas inlet (top) section and break away will possibly block the lower tube rows from where removal will be difficult.

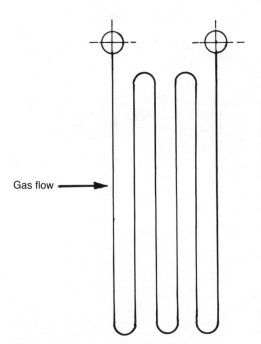

**13.5**  Pendant non-drainable superheater

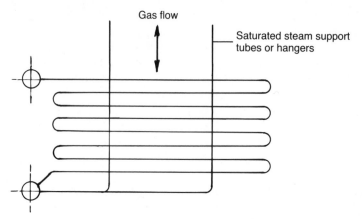

**13.6** Horizontal superheater supported on saturated steam hanger tubes

For fuels with fouling properties, and for systems giving high suspended solids in the flue gases, e.g. spreader stoker firing, the superheater tube pitching across the gas flow should be relatively wide. With normal fuel oils and clean gases, relatively closely pitched tubes can be used with safety, but even so a clear space between tubes of at least 38 mm should be looked for.

Pendant or vertical-tube superheaters of the type shown in Fig. 13.5 are not drainable on the steam side. Drainable vertical-tube designs are

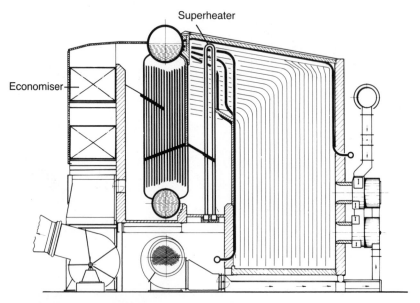

**13.7** Boiler with drainable vertical-tube superheater

possible (see Fig. 13.7) but, depending upon the design of the boiler and the physical size of the superheater, they may be difficult to accommodate within the boiler, and tube replacement can be difficult. The non-drainable type can present an internal corrosion problem if the boiler is shut down for long periods and air is allowed into the steam side of the boiler. Care therefore has to be taken if such a boiler is put into dry storage (see section 6.12.2).

Horizontal-tube superheaters are more easily designed to be drainable. However, should the tubes sag in service, pockets of water will form in the low sections during shut-down when residual steam in the tubes condenses.

Pendant superheaters can be effectively cleared on the gas side on load of all the bonded deposits by the use of long, retractable sootblowers. These can conveniently blow down the gas lanes between the tubes as they traverse the boiler (Fig 13.8). Retractable sootblowers have to be used because the high gas temperature to which they are exposed would overheat the blower tube when not in use and hence uncooled by steam, if it were left in the gas stream. With horizontal superheaters the tubes usually pass across the boiler (Fig. 13.3), so the sootblowers cannot be

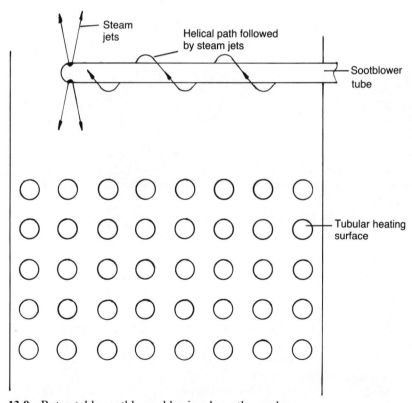

**13.8**  Retractable sootblower blowing down the gas lanes

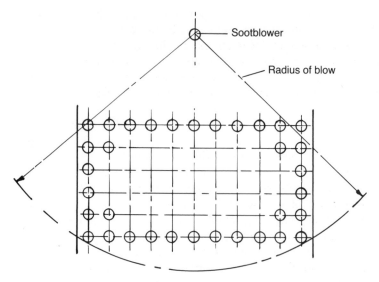

**13.9** Sootblower operating parallel to the axis of the tubes

ideally arranged to blow down all the gas lanes (Fig. 13.9), hence gas-side cleaning may not be fully effective, particularly with deep tube banks, i.e. with a large number of tubes in the direction of gas flow.

When designing superheaters care has to be taken to ensure that the steam flow is equally distributed through all the tubes to give adequate cooling of the tubes and predictable metal temperatures. This is achieved by having a steam-side pressure drop of about 10% of the boiler stop-valve pressure.

To give lower maximum metal temperatures at the superheater steam outlet, thus enabling lower grade materials or thinner tubes to be used, a section of the superheater can be arranged to have steam and gas flow cocurrent (section 5.6).

Boiler start-up can be a difficult period for a superheater, particularly while the boiler water temperature is being raised from cold to the normal atmospheric boiling temperature of 100 °C. During this period there is no steam available to pass through the superheater tubes and so to carry away the heat transferred from the flue gases; the tube design temperatures can therefore be exceeded if the rate of heat input to the boiler is not carefully controlled.

Where a boiler is subjected to frequent off-load periods it can be advantageous to use a higher quality tube material than is theoretically necessary to withstand the normal operating temperatures, so as to give added protection during start-up.

Superheater supports and spacers are important features and require careful design to maintain tube alignment and to allow for differential thermal expansions. Where possible, they should take the form of attachments welded to the tubes to ensure that heat transferred to them from the flue gases is carried away by conduction to the parent tube. Where gas temperatures are high, it may be necessary to use high-quality alloy materials to achieve the desired strength and resistance to scaling. Where possible, tensile stresses should be avoided in high-temperature supports. In many boilers using horizontal superheaters these are supported on tubes carrying saturated steam (see Fig. 13.6 and 2.5), using short, welded attachments between tubes.

## 13.6  Convection evaporative heat-transfer surfaces

When using a multi-drum boiler having a bank of convection heating surfaces of the type illustrated in Fig. 13.3, and there is carryover of particulates in the flue gases, as with solid fuels and some residual fuel oils, care should be taken when selecting the gas flow pattern over the tubes. There is a tendency for particulates to fall from the gases under gravity and at changes of direction of gas flow, thus causing blockage of the gas passages unless removed. Illustrated in Fig. 1.11 is a two-drum boiler which has a cross-baffled convection tube bank; this arrangement is acceptable for clean fuel oils and natural gas. If this arrangement is used with solid fuels there is a possibility of buildup of particulates from the gases above the mud drum due to the change of direction of flow round the lower baffle.

Shop-assembled boilers of the type illustrated in Fig. 2.8, where the heating surface is compact and gives limited access for cleaning, should only be used with clean fuels. If used with fuels producing flue gases containing particulates, these will fall out of suspension under gravity, settle on top of the lower mud drum and eventually restrict the gas flow through the boiler.

The boiler illustrated in Fig. 2.3, suitable for firing coal using a travelling-grate stoker, has been arranged with vertical baffles in the convection bank which are parallel to the axis of the tubes, thus making the gases flow vertically along the tubes. Hoppers have been strategically placed beneath the gas passages at either side of the mud drum, into which the particulates settling out from the gases will fall and from which they can be readily removed.

Large two-drum boilers firing solid fuels are arranged without gas baffles, thus enabling the gases to flow at right angles to the tubes in a single pass. This improves cleaning and reduces particulate fall-out.[4]

When the flue gases have a high particulate loading, low gas velocities must be used to avoid external erosion of the tubes. This is particularly important where baffled tube banks are used which cause the gases to change direction of flow within the tubes. The action of turning the gases concentrates the particulates under the action of centrifugal forces; this creates areas of high dust loading which will severely erode the tubes at a rate which is influenced by abrasiveness of the dust.[4] This action has been frequently experienced on boilers in factories which process sugar cane which are fired with bagasse (cane residue) in which there can be a significant amount of sand.

It is difficult to determine velocities below which erosion will not be experienced because rarely is the gas velocity through such a section of heating surface uniform across and along the gas flow path. Maximum values of about 15 ms$^{-1}$ are quoted for average gas velocities to avoid tube erosion. Tubes on rectangular pitchings are favoured with high dust loadings to assist with cleaning and control of erosion.

The temperature of gas entering and within convection banks is usually sufficiently low to enable fixed multi-nozzle sootblowers to be used. They are arranged with the nozzles positioned to blow down the gas flow lanes between the tubes (see Fig. 7.18). Because steam is emitted from all nozzles at the same time they should ideally be positioned to blow in the direction of gas flow for optimum performance.

## 13.7   Heat-recovery equipment

To achieve an acceptable thermal efficiency it will usually be necessary on all but low-pressure or waste-fired boilers, to include an economiser (section 7.4.1) or airheater (section 7.4.2) to reduce the products of combustion to the required temperature. There are a number of advantages and disadvantages with each, as follows.

### 13.7.1   Economisers

#### 13.7.1.1   Advantages

1.  The tube-metal temperature can readily be maintained at a desired minimum to avoid low-temperature corrosion by feedwater inlet temperature control (Ch. 7).
2.  They give a more compact layout than do airheaters.
3.  They are easier to clean on the gas side than are airheaters, except where the latter are of the tubular type arranged with gas flow across the tubes.
4.  They give a marginally lower heat input to the furnace than airheaters, where the heat recovered is recirculated to the furnace with the combustion air.

### 13.7.1.2 Disadvantage

1. Economisers are pressure parts which, in remote installations having a low standard of feedwater control and operation, can cause an inconvenient outage due to tube corrosion or blockage on the water side.

### 13.7.1.3 Economiser design features

Economisers should, where possible, be arranged with upward water flow. This is to ensure that if the feedwater valve closes due to a high boiler water level, thereby reducing or stopping the water flow through the economiser tubes, any steam generated in the economiser will rise naturally through the tubes due to its buoyancy and be able to escape into the boiler. If this is not done steam pockets can occur which give tube-metal temperatures above the design value.

As indicated in section 7.4.1, economisers can be of the plain or extended-surface tube type. For most liquid and gaseous fuels and normal coals, extended-surface tubes are acceptable, but care needs to be taken with the selection of the fin spacing. When firing fuels which will give high particulate loadings in the flue gases, plain tubes are to be preferred unless the economiser can conveniently be placed after the dust collecting equipment, where the particulate loading will be less. This will mean that the dust collector will be larger due to the higher gas volume to be handled than would have been the case had it been placed after the economiser.

Economisers installed in solid-fuel fired boilers should be fitted with sootblowers for on-load cleaning. These can be of the fixed multi-nozzle or traversing-rake types, the latter being more effective particularly with extended surface tubes (see Fig. 7.20).

It is advantageous, although not necessary, for the economiser to be fitted with a bypass on the water side to allow the boiler to be operated with the economiser out of commission in the event of a tube failure. This necessitates fitting isolating valves to the economiser inlet and outlet pipework to prevent water entering the tubes when the bypass is in use. When fitted with such a water-side bypass it becomes necessary to provide a relief valve to safeguard the economiser tubes against the buildup of excessive pressure should the isolating valves be inadvertently closed while the economiser is in use. A relief valve must also be fitted if there is an isolating or feed-check valve between the economiser and the boiler. The bypass can be used during start-up to enable the tubes to be heated up more rapidly, hence reducing the period of time during which the tubes are below the acid dewpoint of the gases and subjected to corrosion.

## 13.7.2 Airheaters

### 13.7.2.1 Advantages

1. They are non-pressure parts, and hence tube failure is not so serious as with an economiser. Repair is relatively easy without the use of skilled labour such as welders.
2. Hot air can assist combustion of poor-quality or high-moisture fuels.

### 13.7.2.2 Disadvantages

1. They are difficult to protect completely against low-temperature corrosion when firing sulphur-bearing fuels.
2. The performance is susceptible to variations of ambient temperature.
3. They involve the use of ducting to transfer the hot air from the rear of the boiler to the firing equipment; this can impede access around the boiler.

There will be occasions when it is advantageous or necessary to use both airheaters and economisers in combination. Examples are:

1. boilers of the single-drum type where there is little evaporative convection surface; and
2. where hot air is considered necessary for combustion, but the firing equipment imposes a limit on the maximum air temperature, thus making it impossible to achieve the desired final flue gas temperature with an airheater only.

The authors' choice would be an economiser only, unless the combustion of the fuel necessitated the use of hot air.

When firing pulverised coal on boilers which incorporate their own coal milling plant, hot air is necessary for drying-off the surface moisture in the fuel in the pulverisers. In such cases, separate airheaters may be used for the air to the pulveriser to give control of the hot-air temperature.

## 13.8   Constructional features

There are various forms of boiler wall construction still in use, depending upon the boiler type and application, as follows.

### 13.8.1   Brick walls

With thick brick walls having the inner layer formed of refractory bricks,

there may be little or no water-cooling by water-walls on the inside face
(Fig. 1.9 and 1.10). This form of construction is still used on some small
water-tube boilers (Fig. 15.3) and parts of larger boilers (Fig. 15.1) which
fire waste solid fuels in refractory furnaces. The walls can have an external
brick finish or be enclosed in a steel casing.

### 13.8.2   Refractory tile walls (thin walls)

With thin walls consisting of open-spaced water-wall tubes backed by a
layer of refractory followed by one or more layers of insulation, all
enclosed in a steel casing (Fig. 5.1(*a*)), the pressure parts expand within
the refractory and the steel casing (the latter being relatively cool).
Allowance for the differential expansion between the pressure parts and
casing must be made in the design.

### 13.8.3   Tangent tubes

With tangent tubes, i.e. tubes almost touching each other, backed by a
steel casing covered with insulation on the outside (Fig. 5.1(*b*)), the casing
and pressure parts are theoretically at the same temperature, thus
minimising differential expansions. The small spaces between the tubes and
the casing may be filled with castable refractory placed *in situ* before
applying the casing. Pumpable materials are becoming available which can
be injected after assembly of the tubes and casings, thus ensuring a more
satisfactory fill of all the spaces. This prevents the products of combustion
from entering and causing corrosion during shut-down.

### 13.8.4   Welded walls

Welded walls, formed either by tubes welded together with steel strips
between them (Fig. 5.1(*c*)) or by tubes extruded with fins on either side as
an integral part of the tube structure and the fins then welded together,
give a rigid box-like structure for the boiler enclosure. The outside surface
is covered with insulation for heat conservation and personnel protection.
Welded walls reduce the use of refractories, one of the major areas of
boiler maintenance, to an absolute minimum and are now applied to the
majority of the water-tube boilers supplied for oil, gas and coal firing.
Other wall constructions tend to be restricted to small boilers, e.g. those
for special applications or for use in remote areas of low technology.

   Uncooled refractory walls can be favoured as being relatively easy to
repair using unskilled labour, whereas welders of the desired quality may

not be available to repair a tube failure in a welded wall should this happen due to poor operation. Such factors should receive consideration when reviewing available boiler designs and costs for a particular application.

Depending upon the size of the boiler, external steelwork of some form may be required for any type of construction, to provide strength and stability to the walls. The construction referred to in section 13.8.1 (Figs. 1.9 and 1.10) will require a steel structure to support the pressure parts and to give stability to the large areas of brick wall. The construction referred to in section 13.8.2 (Fig. 15.2) will require a structure to support the pressure parts and casings. The constructions referred to in sections 13.8.3 and 13.8.4 (Figs. 2.5 and 2.3, respectively) will require steelwork to give large areas of wall sufficient strength to maintain the wall flat and to avoid overstressing the tubes under the action of the differential pressure across the wall. Steel beams encircling the boiler for this purpose are called girth beams or buckstays.

Water-tube boilers may be either bottom, waist or top supported.

Bottom-supported boilers sit directly on foundations or a supporting structure, with vertical expansion of the hot pressure parts being upward; such a boiler is illustrated in Fig. 1.11. Provision has to be made for horizontal expansion by using sliding supports on the bottom headers and drums.

Top-supported boilers are suspended from an overhead steel structure, with vertical expansion of the hot pressure parts being downwards. Such a boiler is illustrated in Fig. 2.5. Horizontal expansions are more readily catered for by suspending the boiler from sling rods. Where the upper drums or headers sit directly on to steelwork, sliding supports have to be used.

Waist-supported boilers are those that are supported from some intermediate point, either to economise on structural steelwork or to reduce differential expansions between the various parts of the boiler. An example of such a support method is the furnace of the composite boiler which is illustrated in Fig. 1.12, where it would be used to reduce stresses in the connecting pipework between the furnace and the firetube boiler section.

In general, small boilers and two-drum gas- and oil-fired boilers tend to be bottom supported, and all large boilers are top supported. The choice is influenced by the stability of the pressure parts as a bottom-supported structure, and whether or not the boiler is installed in an area subject to seismic disturbances. Bottom supporting tends to give a less expensive construction than top supporting, but heavy lifting equipment may be necessary during construction. The steel structure of a top-supported boiler can be used for lifting the components during construction; it can also be combined into the structure of a boiler-house if an enclosed boiler is required as illustrated by Fig. 2.3.

## 13.9   Internal access

Consideration must be given to access into the various chambers within a
boiler for off-load cleaning, maintenance and repair. It must be realised
that penetrations through the boiler enclosure are potential sources of air
or gas leaks and of heat loss. The latter is particularly important in high-
temperature areas such as access doors into furnaces. Allowing for
observation ports and access through tangent and welded walls will
necessitate the manipulation of the tubes to allow sufficient space between
them, this complicates the wall construction in these areas and of course
introduces the necessity for sealing the openings.

To give access to the furnaces of oil- and gas-fired boilers where cleaning
is unlikely to be necessary between statutory inspection and maintenance
periods, burners can be removed. On solid-fuel fired boilers furnace access
may be through the fuel inlet or ash outlet chambers.

## 13.10   Draught plant

Coal-fired boilers invariably operate under balanced-draught conditions,
with a pressure slightly below atmospheric in the furnace, using both
forced draught and induced draught fans. This is largely due to the
difficulty of sealing against gas leakage around the firing equipment and
fuel inlet with anything other than pulverised coal firing and around the ash
extraction points.

Oil- and gas-fired boilers can operate with forced draught only, that is
pressurised with a positive pressure throughout the air and gas side from
the forced-draught fan through to the chimney.

With modern boiler designs using fully welded walls to enclose the
heating surfaces and furnace, the major areas of the enclosure are readily
made gas tight. There are however still the penetrations through the
enclosure for access doors, observation ports and for superheater tubes and
sootblowers to pass through. Sealing near the drums, and at expansion
joints and dampers requires special consideration. The penetrations for
sootblowers on pressurised boilers have to be provided with cooling and
sealing air. Observation ports which are sealed for pressurised operation
will require cooling air to be supplied to protect the glasses.

The attraction of pressurised operation is to minimise the power
consumption of the draught plant and to eliminate the need for the
induced-draught fan, which is an additional machine requiring attention
and maintenance. The power required to drive a fan is proportional to the
volume of gas or air handled and the pressure generated (section 7.2.6).

With a pressurised system the forced-draught fan generates all the pressure required to overcome the total air- and gas-side resistances from the cold air inlet to the top of the chimney. As the fan handles relatively cold air, the power required is less than that with a system with induced draught plus forced draught. In such cases, the induced-draught fan handles a slightly greater weight of gases at a higher temperature than does the forced-draught fan, and hence handles a considerably higher volume. For example, let: $V$ be the volume of air handled by the forced draught fan; $P_1$ be the system resistance of the air-flow path; $P_2$ be the system resistance of the gas-flow path; the flue gas temperature at the induced-draught fan be 450 K; and the air temperature at the forced-draught fan be 300 K. Then, as in section 7.2.6, for a pressurised system the forced-draught fan power is proportional to

$$V(P_1 + P_2) \tag{13.4}$$

and for a balanced draught system the forced-draught fan power is proportional to

$$V \times P_1$$

The induced-draught fan power (neglecting increased gas weight) is proportional to

$$V \times \frac{450}{300} \times P_2 = V \times 1.5P_2$$

Total power under balanced draught operation is proportional to

$$V(P_1 + 1.5P_2) \tag{13.5}$$

The value of equation [13.5] will therefore be higher than that of equation [13.4].

A complete gas-tight enclosure is difficult to maintain over long periods, particularly with boilers operating at high pressures and where there are frequent off-load periods during which the boiler cools down and causes changing temperature differentials and relative movement of hot parts. Particular problem areas are corners where three planes meet, and interfaces between walls and drums, where welded-wall enclosures cannot be easily connected to the drums. Once a small leak occurs allowing hot gases to escape, the area local to the leak can deteriorate, thus allowing the escaping gas quantity to increase.

It is recommended that, particularly for small industrial water-tube boilers, serious consideration should be given to balanced-draught operation for boilers enclosed in buildings.

The minimum requirements for the design of the pressure parts of water-tube boilers are covered by British Standard 1113 and other equivalent national codes.

# References

1. BS 2885: *Acceptance Tests on Stationary Steam Generators of the Power Station Type*. British Standards Institution, London.
2. BS 845:1987, Part 1 and 2, *Methods for Assessing Thermal Performance of Steam and Hot Water Boilers and Heaters for High Temperature Heat Transfer Fluids*. British Standards Institution, London.
3. ASME-PTC-4-1 1964 Power Test Codes – Steam Generating Units, The American Society of Mechanical Engineers, p. 67.
4. Moir, M. K. and Mason V., 'Tube wear in sugar mill boilers', *Proceedings of Australian Society of Sugar Cane Technologists* 1982 p. 189.

# CHAPTER 14

# Firing appliances

## 14.0　Introduction

The physical principles upon which firing appliances depend and on which
they should be designed have been discussed in Chapter 4. The purpose of
the present chapter is to describe some typical firing appliances for solid,
liquid and gaseous fuels. The design of firing appliances used in firetube
boilers, will, because of size, be different from those used in water-tube
boilers, although the principles will be largely similar. The space available
in the furnaces of firetube boilers for the accommodation of firing
appliances is much more restricted than in water-tube boilers. The volume
available for combustion is also much less. These factors combine with
overall size to dictate the differences in construction.

This chapter commences with the smaller, and generally simpler firetube
boilers, for which established types of appliances for coal, oil and gas will
be described. This will be followed by a similar treatment for water-tube
boiler firing appliances. In both cases, examples of fluidised-bed
combustion appliances will be included, although at this period of history it
cannot be said that this system is in common use, being still in the
development stage.

## 14.1　Firing appliances for firetube boilers

### 14.1.1　Coal firing

As has been explained in Chapter 4, there are three principles of
combustion, any one of which may be used; these are overfeed, underfeed
and composite. Which principle is used depends on the kind of coal
available, particularly the size distribution, and the rank, which latter is a

measure of its caking properties. Because of the dimensional limitations of
firetube boilers, washed coals only, with ash contents of less than 10%, are
generally used; larger and more sophisticated water-tube boilers may use
untreated small coals, particularly when pulverised, with ash contents up to
20% or even higher, although economic factors may determine otherwise
at the smaller end of the range.

### 14.1.1.1  Overfeed combustion

Figure 14.1 illustrates the practical application of the principle illustrated in
Fig. 4.6, which is a firetube boiler in which singles coal is supplied to a
'drop tube' through which the coal is dropped vertically downwards on to a
fixed grate. More or less even distribution of the coal over the grate is
achieved by imparting rotary motion to the coal by air admitted
tangentially to the feeder, this motion being transferred to the coal. A
coned discharge helps to distribute the coal, and the conveying air acts as
secondary air to complete the over-fire combustion.

It will be appreciated that this method of feed will not result in a
perfectly uniform bed, since the coal delivery pattern will be circular,
whereas the grate is rectangular; the fuel feed will therefore vary in
thickness from place to place. To overcome the passage of too much air
through the thin places, the grate is designed to present a large resistance
to air flow in relation to that presented by the fuel bed. The air flow
through the grate and fuel bed combination will be more or less uniform.
Dead plates prevent air flowing through those parts of the fuel bed which
cannot be covered adequately with coal. Given a carefully graded fuel,
singles, with very little fines, the performance of this type of firing
appliance can be excellent, but if fines are present, difficulties can arise
from their entertainment in the over-fire gases and their carriage forward
through the boiler. The thermal loss due to heavy carryover can be
significant, and the grit can also erode the inlets to the boiler tubes. Grit
from the grit arrestor at the rear of the boiler can be refired, but this
increases the burden in the boiler gases and hence the erosive effect
(section 16.2.1).

The advantages of this system are compactness and, including the coal
handling system to the boiler, cheapness. Some disadvantages of the
system have been mentioned in the previous paragraph, another is the
need for manual attention. A mechanical grate has now been developed,
sections of which dump parts of the fuel bed in sequence into the space
beneath. Ash is removed by a conveyor to an ash disposal system at the
front of the boiler. The bars on which the grate sections rotate are water
cooled.

It will be seen that this method of firing is an integral part of the boiler
and is inseparable from it. Another system of overfeed firing, which is
applicable to any firetube boiler, is illustrated in Fig. 14.2. The coal is

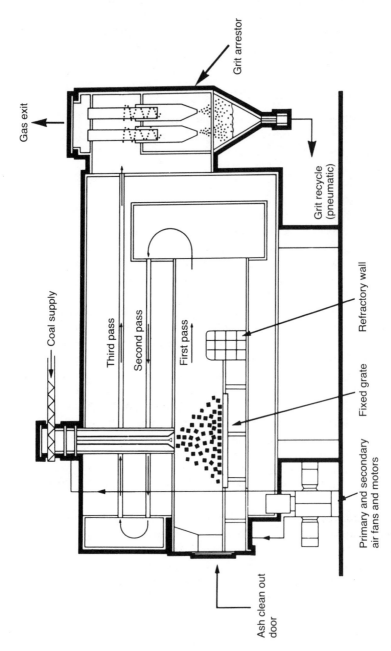

**14.1** Firetube boiler with vertical drop tube and fixed grate

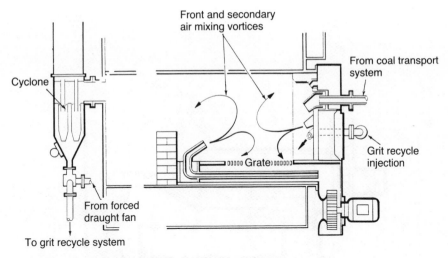

Cyclone

Front and secondary
air mixing vortices

From coal transport
system

Grit recycle
injection

000000 Grate) 000000

From forced
draught fan

To grit recycle system

**14.2**  Firetube boiler with fixed grate and horizontal coal injection
(From Saacke Limited)

conveyed by lean-phase transport to the front of the boiler, whence it is
pneumatically projected over the grate; adjustable splitters are provided to
assist even distribution. Manual de-ashing is used: the ash is drawn towards
the front of the boiler into a purpose-designed hopper which can be
wheeled away; or the ash is discharged into a conveyor system. The same
general remarks apply to this system as to that described previously,
particularly with regard to the quality of coal that can be used. With both
types, it is particularly important that the coal, which may be delivered as
good quality singles, is not degraded in handling between storage and
boiler. This can easily happen if excessive air is used to transport the coal
in a lean-phase system; screw conveyors can also cause degradation. Coals
with low friability are obviously necessary for this kind of firing, and great
care is necessary in their selection.

### 14.1.1.2  *Underfeed combustion*

Figure 14.3 shows a section through a firetube boiler fitted with a chain
grate. This is the practical application of the principle illustrated in Chapter
4, Fig. 4.7. The fuel bed is moved gently from the front of the boiler to the
rear of the grate, whence ash, clinker and any unburned fuel, which last
should be small in quantity, are discharged through an ash drop tube,
either into a purpose designed barrow or into a conveyor system. In either
case, it is important that air ingress into the drop tube is prevented,
otherwise combustion conditions will be upset and fine ash particles will be
entrained and carried forward to cause problems elsewhere in the boiler.
An earlier method of ash discharge, before the innovation of the drop

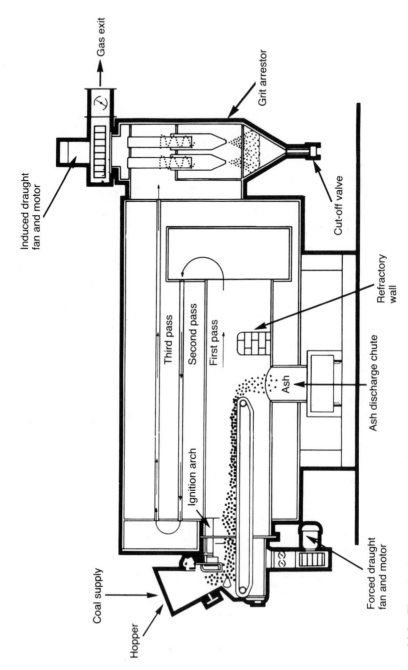

Gas exit

Induced draught fan and motor

Grit arrestor

Cut-off valve

Third pass

Second pass

First pass

Refractory wall

Ash

Ash discharge chute

Ignition arch

Coal supply

Hopper

Forced draught fan and motor

**14.3** Firetube boiler with chaingrate

tube, was to draw the ash forward, either manually or mechanically, to the front end of the boiler, whence it was removed by barrow or discharged into a conveyor system.

Chain grates work well with small coal[1,2,3], but as mentioned in Chapter 4, it is essential that the coal is properly conditioned with moisture. Small washed coal is often wet when delivered, but may dry out during storage. Moisture on the particle surfaces causes mutual adhesion leaving interstices through which air can penetrate the bed and react with the coal. If insufficient surface moisture is present, fines can clog these interstices preventing air ingress, and this interferes with combustion (see Ch. 4).

With chain grates, ash content is important and should normally exceed 5%, otherwise protection of the grate, particularly at the rear end, is insufficient to prevent the grate overheating due to exposure at reduced load to flame radiation. Singles can be difficult to burn on chain grates owing to the slow ignition rate inherent in underfeed combustion. Coal particles ignite at their corners, and the larger the particles, the fewer are the corners and the slower will be the ignition rate. Well-graded singles, desirable for overfeed combustion, may be difficult to ignite on chain grates owing to the reduced number of corners compared with smalls.

### 14.1.1.3   Composite combustion

Figure 14.4 illustrates a coking stoker installed in a firetube boiler. This is similar to a chain grate in that the grate moves the coal from front to rear, discharging ash at the rear whence it can be removed by a drop tube or from the front of the boiler. The ignition rate is faster than on a chain grate because the fresh coal is injected into the fuel bed between two ignition planes (see Fig. 4.8). The coking stoker is therefore very suitable for burning singles and has advantage over the pure overfeed systems described in section 14.1.1.1 of this chapter in that the coal is not airborne before reaching the grate. This firing appliance is therefore much less sensitive to the fines content of the coal than is the pure overfeed type previously described, and is indeed suitable for burning many washed smalls coals provided these are not of the caking type.[4,5] Chain grates will tolerate strongly caking coals better than coking stokers will since the coal is moved more gently on a moving chain than on reciprocating bars which tend to compress the coal on their return stroke.

### 14.1.2   Oil firing

#### 14.1.2.1   Pre-burner requirements

BS2869:1983 specifies fuel oil for oil engines and burners for non-marine use. The properties of oils are tabulated under a number of class headings,

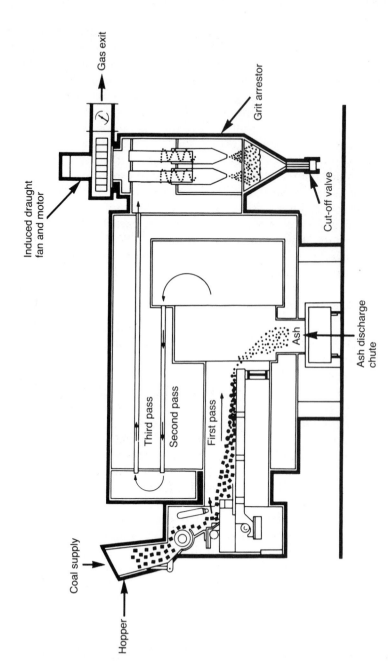

**14.4** Firetube boiler with mini coking stoker

and it is class G, often known as 'heavy fuel oil', which is mostly used in industrial boilers, although class D, 'gas oil', is sometimes used as a stand-by when natural gas is the main fuel. Most oil burners are therefore designed for using heavy fuel oil, which must be heated to about 85–135 °C, according to the method used for atomising the oil, for satisfactory combustion. Heavy fuel oil is delivered at a temperature of around 60 °C and must be stored at a temperature which will allow the oil to flow to the pumps supplying it to the burners. The minimum storage temperature recommended for this grade of oil is 40 °C, after which it must be raised in pressure and temperature suitable for delivery to the burners. This is done in a pumping and heating set which circulates heated oil to the oil-using equipment, any surplus being returned to the pump inlet or storage tank. Filtration is also needed to remove foreign matter, often fibrous in nature, which would otherwise clog the burner nozzles. Further heating and increase of pressure to precisely controlled conditions may be performed on the burner itself.

It will be seen therefore that, before the oil is atomised, mixed with air and burned, there needs to be a complex hydraulic system for preparing the oil for this purpose. This section is concerned with the burner itself rather than the storage, preparation and distribution of the oil. It will be realised that with gas oil, because of its much lower viscosity, six centistokes at 40 °C, compared with heavy fuel oil which is up to 85 centistokes a 100 °C, no heating will be required, thus greatly simplifying the preparation of the oil. (The limit for heavy fuel oil will be altered to 85 centistokes at 100 °C in the revision of BS2869.) Gas oil, however, is too expensive for general use, but the simplification is often attractive when a stand-by fuel is needed.

### 14.1.2.2   Atomisation

**14.1.2.2.1   Pressure-jet atomisation**   Before fuel can be mixed with and burned with air it must be atomised into a fine spray. This can be done by raising the oil pressure and discharging the oil tangentially into a circular chamber, the 'swirl' chamber. It issues from this through an axial hole as a rapidly rotating film, this film then breaking up into small particles. The included angle of discharge, often about 70°, is achieved by carefully machining the discharge orifice to the required shape. It is important that the surface finish of the atomiser is of very high quality, otherwise small defects, scratches for instance, will cause premature breakdown of the rotating film and the production of coarse droplets which are difficult to burn. To achieve the high velocity needed for good atomisation, pressures in the range 20–40 bar are required. Atomisation tends to be very poor below 10 bar. Since flow rate varies with the square root of pressure it follows that, if the pressure corresponding with maximum flow rate is

40 bar, the minimum flow rate at which satisfactory atomisation can be achieved will be

$$\left(\frac{10}{40}\right)^{0.5} \times \text{maximum flow rate} = 0.5 \times \text{maximum flow rate}$$

This limits the turndown ratio of a pressure jet burner to 2:1 which is generally not sufficient for industrial boiler operation.

To overcome this limitation, multiple nozzles may be used, but this tends to interfere with the symmetry of mixing (see Ch. 4, section 4.1.5) and hence with good combustion. Another method is to maintain the angular velocity in the swirl chamber at all loads by bleeding off a proportion of the flow from the swirl chamber back to the inlet of the burner pump. This is known as the 'spill jet' system. It has however a disadvantage in that the angle of the spray varies with the turn down, again interfering with good combustion. Figure 14.5 illustrates diagrammatically the circuit of a spill return pressure-jet burner, and Fig. 14.6 shows a section through a commercial burner of this type for a small firetube boiler (see also Fig. 4.4). Fig. 14.7 illustrates a spill return burner, arranged also for natural gas firing, of the type used on a water-tube boiler.

Over the past few years there has been an increasing demand for the premium grades of oil, gasoline, derv, and aviation fuel. This has meant that more of the highly viscous components of the crude oil are concentrated in the residual heavy fuel oil which is becoming increasingly more difficult to atomise. Before that time, pressure-jet atomisers could handle heavy fuel oil with considerable ease, accepting their limited turn-down ratios. The more recent fuel oils, however, present difficulties with the process of atomisation, which cannot now be considered as satisfactory, except for the lighter oils, i.e. medium fuel oil (BS2869 class F) and below.

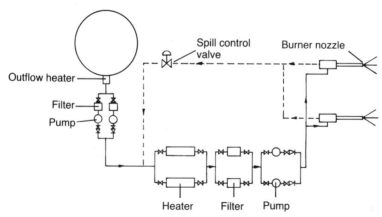

**14.5** Hydraulic circuit for a pressure jet oil burner

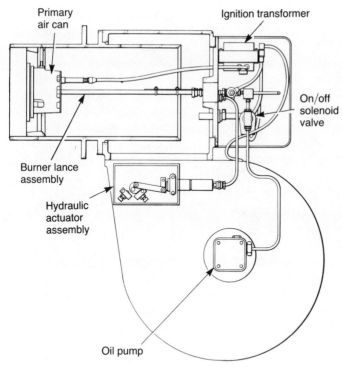

Primary
air can

Ignition transformer

On/off
solenoid
valve

Burner lance
assembly

Hydraulic
actuator
assembly

Oil pump

**14.6**   Commercial pressure-jet burner for a firetube boiler

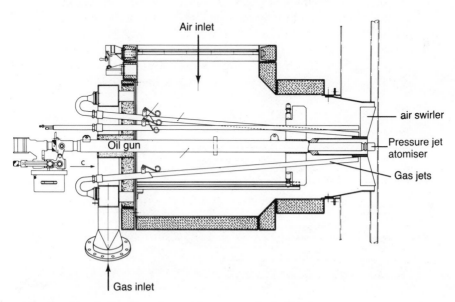

Air inlet

air swirler

Oil gun

Pressure jet
atomiser

c

Gas jets

Gas inlet

**14.7**   Combined oil and gas burner for water-tube boilers

**14.1.2.2.2 Twin fluid atomisers** The object of these is to overcome the shortcomings of pressure-jet atomisers by using a secondary fluid, air or steam, to provide the high velocity needed to atomise the mixture. The velocity and quantity of the secondary fluid is independent of that of the oil, so that the quality of atomisation is much less affected by reducing the rate of flow of oil than with pressure jets. Turn-down ratios of 4:1 or better are achieved, which fill the requirements of most industrial boilers. Moreover, the high pressures needed for pressure-jet atomisation are not now required. Twin-fluid atomisers are now widely used in industrial boilers and can provide satisfactory performance with current qualities of heavy fuel oil.

There are two main types of twin-fluid atomisers.

1.  The first type comprises those in which the oil and atomising fluid are mixed in a multiplicity of nozzles when they are ejected in finely divided sprays which combine to form a single conical spray with an included angle of around 70°. The secondary fluid, air or steam, is provided at a pressure of between 1.4 and 8 bar, the former being referred to as 'medium-pressure air' and the latter as 'high-pressure air or steam'. These types of atomisers are generally used in water-tube boilers. Operational costs tend to be high in the case of steam atomisation due to the cost of the extra treated water and heat loss. In the case of air atomisation, the cost of compressed air must be considered.

    Figure 14.8 shows a typical air or steam atomiser nozzle.
2.  The second type comprises those in which oil is spread over the internal surface of a rapidly rotating (5000 rev min$^{-1}$) conical cup machined from stainless steel. The oil is discharged from the wide end of this as a thin radial film, which is broken up and atomised by a peripheral high-velocity air jet from a fan discharging at about 44 mbar. This is known as the 'rotary cup burner' and is widely used

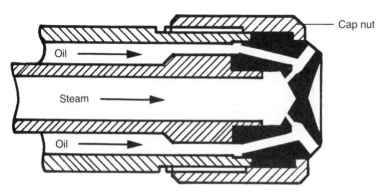

**14.8** Air or steam atomiser
(From Saacke Limited)

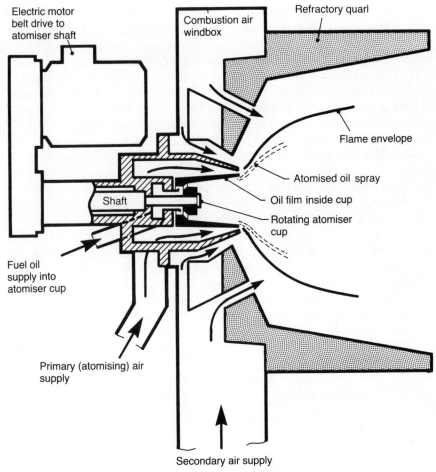

Electric motor
belt drive to
atomiser shaft

Combustion air
windbox

Refractory quarl

Flame envelope

Atomised oil spray

Oil film inside cup

Shaft

Rotating atomiser
cup

Fuel oil
supply into
atomiser cup

Primary (atomising) air
supply

Secondary air supply

**14.9**   Principle of a rotary cup burner
(From Hamworthy Engineering Limited)

for firetube boilers. Figure 14.9 illustrates a simplified section through a burner of this type. Fraser, Dombrowski and Routley[6] have studied in detail the parameters which affect the performance of rotary cup burners, much of their work having now been embodied in current designs.

Rotary cup burners, because of the mechanical effort used in atomisation, require air and oil at relatively low pressures and can also tolerate higher oil viscosity, thus requiring lower oil temperatures than most other types of burner. Typically, a temperature of 85–90 °C is adequate for heavy fuel oils, although with the current changes in oil refinery practices this may tend to rise slightly in the future.

The primary air used for atomisation and penetrating to the heart of the flame is some 14% of the total quantity. Most of the air is secondary air, supplied through radial slots directed axially to the furnace but converging in the flame zone from a pressure of about 20 mbar (see Fig. 14.9). A small quantity of tertiary air is admitted, the main purpose of which is to keep the air ducting which surrounds the cup at an acceptably low temperature.

Firetube and some water-tube boilers fired by oil or gas operate with a positive furnace pressure of 10 mbar or more in order to drive the products of combustion through the boiler passes and through following equipment such as economisers, ducting and the chimneys. The forced-draught fans therefore need to cater for the back-pressure exerted by the particular boiler design and by that of the subsequent equipment. Some margin in pressure and volume delivery of the fan is needed to overcome fouling of the boiler during use (section 7.2), but this should be minimal, otherwise over-loading of the boiler can occur.

It is important that oil shall not drip into the furnace of the boiler when combustion is not taking place, otherwise fractions can distill off in the hot environment and accumulate. An explosion could occur when the burner is next ignited. This is particularly hazardous when gas oil is used, as ignitable fractions occur at lower temperatures than with residual oils. In order to overcome this danger, it is important that the hydraulic system is provided with duplicate shut-off valves which are 100% reliable, particularly with gas oil. It is also required that the furnace is purged with air immediately after shut-down and again before the ignition source is actuated; BS799[7]:Part 4:1972 and Part 6:1979 are relevant to these matters. BS5410[8]:Part 2:1978 applies to oil burning systems in general. All these standards should be studied.

## 14.1.3   Gas burners

Natural gas is the easiest of all fuels to burn, as no fuel bed is required, and neither is atomisation; all that is required is to mix one gas, the fuel, with another gas, air, in metered proportions and with proper mutual distribution. Natural gas for firing industrial boilers is mostly supplied on an interruptible tariff, oil being used as the standby fuel. Thus dual-fuel burners are more commonly used than those designed for gas only, and this facility has been well developed by the burner makers.

Basically, a gas burner for a boiler is very similar to an oil burner, consisting of a fan (or fans) to supply the necessary air at the appropriate pressure, and gas discharge apertures to distribute the gas into the air. Figure 14.7 shows a combined oil and gas burner with the flame stabilised by a bluff body. The gas discharges from a number of tubes the exits of which are terminated just behind the bluff body and near its edge.

Figure 14.10 shows a dual-fuel rotary cup burner. It will be seen that this is very similar to the oil burner shown in Fig. 14.9. Gas is admitted to a

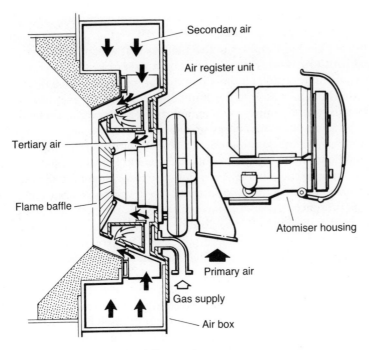

**14.10**  Dual-fuel burner (oil or gas)
      (From Saacke Limited)

distribution chamber which is contained within the chamber from which air
is delivered to the secondary air slots. Gas from the gas chamber is
admitted into each secondary air slot, thus giving a good final distribution.
The amount of gas is regulated by a modulating valve which is supplied
from a governed, constant-pressure source.

As with oil burners, it is important that under no circumstances shall gas
enter the furnace unless it is intended to do so. The consequences of an
explosion with gas are likely to be even more violent than those with gas
oil. To ensure this, two positive shut-off valves are used in series in the gas
supply to each burner. The pipe connecting these is purged with nitrogen
to obviate the passage of gas leaked by a faulty first valve into the second.
Alternatively, the cavity between the two valves is vented to atmosphere
when they are closed, this is known as a double block and vent system.
These valves are electrically operated and connected with the boiler
control system. The gas industry in the UK has been greatly involved in the
design and testing of such arrangements, which, when properly installed,
can be considered to be 100% safe. BS5963[9]:1981 specifies these shut off
valves whilst BS5885[10]:1980 deals with industrial gas burners of the types
used for firing industrial boilers.

## 14.1.4 Firing equipment for industrial boilers: a general comment

The introduction of 'package' boilers in the 1950s (see Ch. 1) and the trend towards oil and then to gas firing meant that as old boilers were replaced, new boilers became almost fully automatic in operation. Perhaps the biggest step in this direction was the development of automatic ignition, enabling a boiler to start up entirely unmanned, from a signal from a clock or, in more recent years, from an energy management system.

In pre-war years automatic ignition was developed for domestic and industrial gas appliances, first by the use of a permanent gas pilot flame. Should the pilot fail gas would be delivered into the environment with great risk of explosion. It was therefore essential to develop devices which would detect the presence of a pilot flame and to ensure that the gas supply was cut off automatically should this be absent. For the larger appliances, where a permanent pilot could be wasteful, the use of electric spark ignition of the pilot flame, which then ignited the main flame became established.

With the development of oil burners for industrial boilers, it was a short step to arrange for the electric ignition of a gas pilot or, for smaller burners using distillate fuel, for the direct ignition of the atomised oil by a high tension spark. The need for safe operation of these devices is obvious, the absence of a spark, of an adequate pilot flame or of a main flame must be detected and the fuel supply cut off. Much of the sophistication of modern burners is concerned with reliable flame detection and subsequent lock-out should the required spark and flame not be proven. It is here that references 7, 8 and 9 are of great importance. The flow of sufficient air to the burner must also be proven, otherwise the burner can run through its start-up sequence; there may be sufficient air only to support a pilot, and when there is an attempt to ignite a main flame lock-out occurs. A gas or oil burner is therefore quite a complex device.

The recent trend back to coal firing and the need to compete successfully with oil or gas firing has meant that the safe and reliable ignition of fuel beds has had to be studied. This is not at all easy, although it is entirely feasible with pulverised fuel and with fluidised beds, both of which have been developed with automatic ignition as a necessity. With moving grates, chain grates and coking stokers, approaches have been made using gas pilots or air, electrically heated to around 700 °C. Whilst these devices can work, reliable flame detection and protection, on the scale achieved with oil or gas firing, is difficult as initial and local production of smoke tends to interfere with the detecting apparatus. At the time of writing the normal procedure is manual ignition using a portable gas torch or even the old method of 'sticks and paper'.

## 14.2   Firing appliances or combustion equipment for water-tube boilers

Water-tube boilers do not suffer the restraints to furnace size imposed on firetube boilers by the restrictions on furnace dimensions, and hence have few restraints on the type of combustion equipment that can be applied to them. As a consequence, there is much more flexibility with the range of fuels that can be fired and the shape and rating of the furnace that can be used. Water-tube boilers can utilise all the types of solid fuel firing appliance used on firetube boilers as well as those designed for pulverised fuel firing and for the combustion of bulky biomass fuels. This section of the chapter indicates the most common types of combustion equipment currently used in water-tube boilers and mentions areas where development is being carried out.

While hand-fired grates are still used on small boilers in low-technology areas, combustion is inefficient due largely to poor air distribution through the fuel bed and consequent high excess air values, and their use is therefore limited.

### 14.2.1   Stokers

These operate on the 'underfeed' and 'overfeed' principles as defined in Chapter 4.

#### 14.2.1.1   *Stokers using the underfeed principle*

Stokers using the underfeed principle are of the travelling grate type, and there are two basic types in use.

1.  The chain grate stoker has the same general construction as that used with the firetube boiler (see Fig. 14.11). This consists of a broad endless belt of castings in which each row of links across the stoker is connected to the adjacent row by steel rods. Special links provide the drive at the sides of the grate and at intermediate locations.

    Permanent replacement of the grate links necessitates removal of the steel rods holding the links, and special links are available for short-term repairs.

2.  The travelling grate, like the chain grate, has a number of separate driving chains across its width on to which the grate bars, or louvres, and their rods are mounted. They are attached in such a way that as the grate reaches the rear ash discharge end, the bars hang downwards allowing any ash or coal particles held between them to fall out into the ash hopper, giving a self-cleaning feature (Fig. 14.12). The grate

**14.11** Chain grate stoker for a firetube boiler

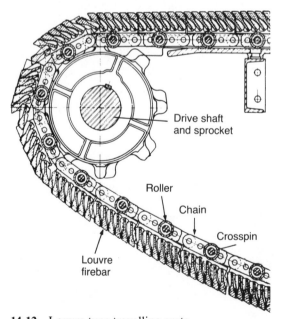

**14.12** Louvre type travelling grate

bars are relatively easy to replace, and this can be done through doors at the front of the stoker while the boiler is on-load.

Unlike the firetube chain grate, there is usually an arch over the rear end of the grate. The purpose of this is to maintain sufficiently high temperatures in this region to ensure that most, if not all, of the coke in the ash is completely burned. In the firetube boiler the flame passes over the rear end of the grate and therefore performs this function.

With water-tube boilers there are fewer constraints on the depth of the stoker than on a firetube boiler, and the stoker is accessible all round for the introduction of combustion air. As a consequence, it is possible to achieve precise control of the undergrate air by the use of dampered compartments, as illustrated by Fig. 4.10, or of air distribution dampers, as illustrated by Fig. 14.13.

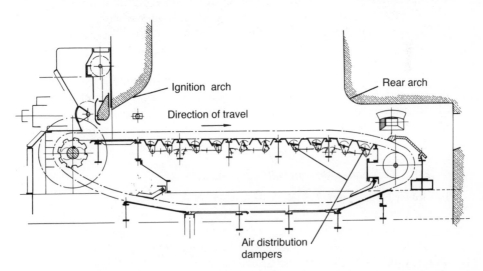

**14.13**   Travelling grate stoker for a water-tube boiler

In both of the foregoing cases the grate travels from the front of the boiler to the rear and is driven by the front shaft, giving a continuous ash discharge facility. The fuel is fed on to the grate from an overhead bunker through a chute, which may be of the traversing type (Fig. 2.3), into a hopper at the boiler front. The depth of the fuel bed is regulated by a guillotine which is raised or lowered to suit. The rate of fuel feed to the furnace is controlled by a combination of grate speed and fuel bed depth, the best combination to suit the fuel used being determined by operation.

The grate drive is usually continuous through a variable-speed gearbox, and grate speeds are up to six metres per hour. Front and rear overgrate secondary air is provided in the form of high-velocity air jets to ensure adequate mixing of the combustion air and volatiles driven off from the fuel bed; this gives completely smoke free combustion.

The moving link type of stoker can burn a very wide range of coals, but the correct combination of grate rating and front and rear arch above the grate have to be selected to give early ignition of the incoming fuel and complete burn-off at the rear.

### 14.2.1.2   Spreader stokers with travelling grates

Overfeed or spreader stokers for coal firing use rotary mechanical spreaders to distribute the coal along the furnace, spreading it over the grate surface. As the larger pieces of coal travel further towards the rear of the furnace, and hence the grate, than small pieces, the grate surface travels from the rear of the furnace to the front, giving the larger pieces a longer residence time on the grate to ensure complete combustion. Figure 15.2 illustrates a boiler with a travelling-grate spreader stoker.

The grate construction can be the same as that used for the travelling-grate stoker or have a few long grate bars across the boiler width; these are again fixed to separate driving chains. The term 'continuous ash discharge' (CAD) spreader stoker is among those used to differentiate it from the dumping grate spreader. The ash is continuously discharged from the front of the grate instead of at the rear, as with the underfeed travelling grate. Drives can be continuous through a variable speed gearbox or intermittent using hydraulic cylinders.

### 14.2.1.3   Spreader stokers with dumping grates

For low ash fuels, particularly fibrous fuels (Ch. 15), dumping grates can be used instead of continuous chains. The grate surface consists of rows of grate bars about 300 mm wide which can be tilted by mechanical linkage using manual or mechanical actuators (see Fig. 14.14). Combustion is less efficient than with the continuous ash discharge type due to the poor size distribution of the fuel across the grate and the fact that unburned fuel is discharged when dumping. The grate is sectionalised to enable one section to be dumped at a time, hence maintaining burning fuel on the remaining grate sections to reignite the fuel on the dumped section when it is closed. The rate of fuel feed to the furnace with spreader stokers is controlled by the use of variable-speed feeders of either drag link, pusher or rotary drum type (of which Fig. 14.15 is typical) discharging the fuel on to the rotating spreader.

### 14.2.1.4   Characteristics of spreader stokers

With spreader stokers a proportion of the fuel burns in suspension; this can be as much as 50% depending upon the fuel grading and density. In the case of fibrous fuels such as bagasse and wood, where the fuel density is

Side castings

Tilting grate bars

Grate
support bars

**14.14**  Dumping grate stoker

low compared with coal, the percentage burning in suspension can be expected to be higher than with dense coals. Spreader stokers are very versatile and will burn all but low-volatile coals. Low ash contents can result in the exposure and overheating of the grate surface. The response to load changes of a boiler fired with a spreader stoker is good. Overgrate secondary air is applied and is probably more important than on an underfeed stoker due to the high quantity of suspension burning.

### 14.2.1.5  *Stoker ratings and fuel grading*

Stoker dimensions are determined on the basis of heat rating per unit of effective surface area ($MW\ m^{-2}\ s^{-1}$); see Table 14.1. The ratings at which CAD spreader stokers can operate are higher than those with underfeed travelling grates by a factor of about 1.5, due to the fuel burning in suspension. A consequence of this is that the plan area of the grate of a CAD spreader stoker is less than that of a travelling grate one for a given boiler output and fuel quality. Because the residence time for satisfactory fuel burn-out and hence volume release rate is at least the same as for an underfeed travelling grate, the furnace height will be greater than with an underfeed travelling grate to give the equivalent furnace volume.

The maximum output available from a boiler using travelling grates is

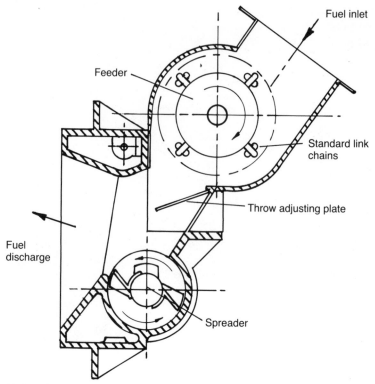

**14.15** Rotary feeder and spreader

**14.1** Typical heat loadings for stoker effective areas (coal firing)

| | |
|---|---|
| Underfeed travelling grate | 1.1–1.42 MW m$^{-2}$ hr$^{-1}$ |
| Continuous ash discharge spreader | 1.9–2.37 MW m$^{-2}$ hr$^{-1}$ |
| Dumping grate spreader | up to 1.42 MW m$^{-2}$ h$^{-1}$ |
| Pulverised fuel (furnace plan area) | up to 4.7 MW m$^{-2}$ h$^{-1}$ |

limited by the maximum grate size available, and this is dictated by mechanical considerations such as size of drive shaft. The maximum dimensions of a travelling grate, including CAD spreaders, is about 11 m wide by 8 m long using two grates side by side. The maximum boiler output is about 28 kg s$^{-1}$ (78 MW) for underfeed travelling grate and 42 kg s$^{-1}$ (177 MW) for a CAD spreader stoker, depending upon the steam conditions and fuel quality.

Because the fuel is thrown into the furnace on a spreader stoker, there is a higher carryover of fuel and ash particles with the flue gases into the boiler convection heating surface than is experienced with the fuel bed of an underfeed travelling grate. This results in a reduced efficiency due to the higher unburnt fuel loss. This is compensated for, to a certain extent, by refiring some of the larger particles (section 13.2.1.4).

Graded coals are preferred for spreader stokers to improve distribution over the grate surface. The underfeed travelling grate can handle smaller coal than can the spreader.

### 14.2.1.6   Other types of stoker

Retort or reciprocating grate stokers are used for fuels where fuel bed agitation is essential to ensure mixing of the fuel and combustion air on the grate. A typical example of the use of this type of grate is for mass burning of municipal refuse (section 15.1.2).

For fibrous fuel there are a number of alternative grates in use, consisting of inclined finned tubes with air ports in the fins. These can be static or vibrating, the latter giving continuous ash discharge facilities. Slightly inclined furnace floor tubes covered with perforated castings and referred to as 'pinhole grates' are used. Air or steam jets are arranged just above the grate surface to blow the ash towards the lower end of the grate for removal.

### 14.2.2   Pulverised fuel firing

When the output of a coal-fired boiler exceeds the capability of the maximum size of stoker or with high fines or high ash coals, pulverised coal firing is used. Most utility boilers are now of such a size that pulverised fuel firing is now almost universally used on coal-fired boilers. With this method of firing the fuel is ground until 90–95% will pass through a 200 mesh sieve (74 $\mu$m), i.e. it has the consistency of flour.

The thermal efficiency of a boiler fired with pulverised fuel is higher than that of a boiler fired with a stoker for a given fuel and final flue gas temperature. This is largely brought about by a lower heat loss due to incomplete combustion of the fuel, the unburnt carbon content of the ash and grits being lower than with stoker firing. Some improvement in efficiency also results from the ability to reduce excess air quantities for combustion, this being possible because good mixing of fuel and air is more readily achieved in burners with the very fine fuel which exposes a much greater surface area to the air than does the fuel on a stoker.

Grinding of the fuel is carried out in mills or pulverisers. There are a variety of designs, the major ones being the following.

1.   In attritor mills the fuel is broken up by impaction with rotating studs or hammers.
2.   In ball mills the fuel is ground in a cylinder containing hard metal balls, rotating with its longitudinal axis horizontal.

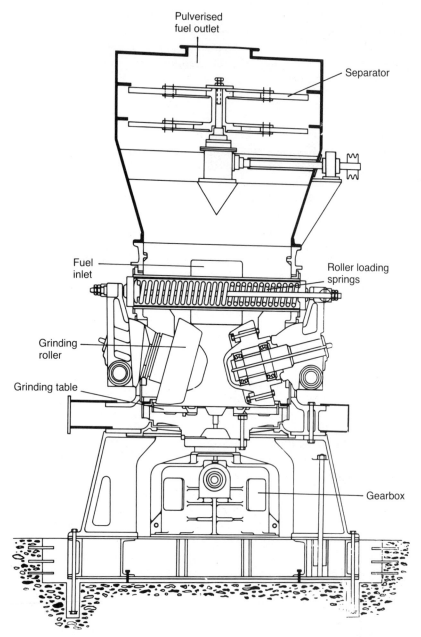

Pulverised
fuel outlet

Separator

Fuel
inlet

Roller loading
springs

Grinding
roller

Grinding table

Gearbox

**14.16** Typical roller and table pulverising mill

3. In roller mills the fuel is ground between rollers and either fixed or rotating tables or bowls. A typical design is illustrated by Fig. 14.16.
4. In pneumatic or steam jet mills particles are broken down to a very

small size by mutual attrition in highly turbulent conditions (see section 14.5.1).

The mills can operate under a positive or negative (suction) pressure.

Pressurised mills operate with clean air fans but require tight sealing to prevent the loss of fuel to the atmosphere and thus causing a nuisance and explosion hazard. Suction mills use exhauster fans which handle the fuel and air mixture from the mill and are subjected to erosion, thus requiring frequent maintenance.

The fuel can be pulverised in central depots for transportation to the various boiler plants. The more widely used system for large boilers is to pulverise the fuel at the boiler front and to convey it direct from the pulveriser to the furnace with the primary combustion or carrying air. Hot air is invariably supplied to the mills, the heat content being sufficient to evaporate the surface moisture and thus improve the grinding process.

The fuel can be fired in one wall only, usually the front wall, or be corner fired (Fig. 13.4). This latter system is capable of firing coals of lower volatile content than is front-wall firing largely because the concentration of heat from all burners into one flame core helps to drive off the volatiles and ignite the carbon.

For a given furnace heat input the furnace of a pulverised-fuel fired boiler will have a smaller plan area than a stoker fired one (Table 14.1), but as a consequence will be much higher in order to achieve the desired furnace residence time and burn-out of the fuel. This tends to give a less expensive boiler than for stoker firing but, taking into account the cost of the grinding mills and fuel feeding system, the higher power consumption and the requirement for a more efficient gas cleaning plant due to the fine dust particles contained in the flue gases, the overall cost is higher. Pulverised-fuel firing requires a more sophisticated control and combustion management system than does a stoker fired boiler.

### 14.2.3 Fluidised bed combustion

Fluidised bed combustion (FBC) is being developed for both firetube and water-tube boilers. When such a bed is used in the cylindrical furnaces of horizontal firetube boilers (Fig. 14.17), the elutriated material cannot fall back into the bed and is carried forward horizontally into other parts of the boiler, whence it has to be extracted and returned to the bed. The furnaces of water-tube boilers can be made to any size and shape, giving much more freeboard above the bed and thus allowing a considerable proportion of the material carried from the bed to return naturally by gravity. Some material will, however, be carried forward to collect in other parts of the boiler and will need to be returned to the bed, but this is very much less than with firetube furnaces. Water-tube furnaces are therefore much more

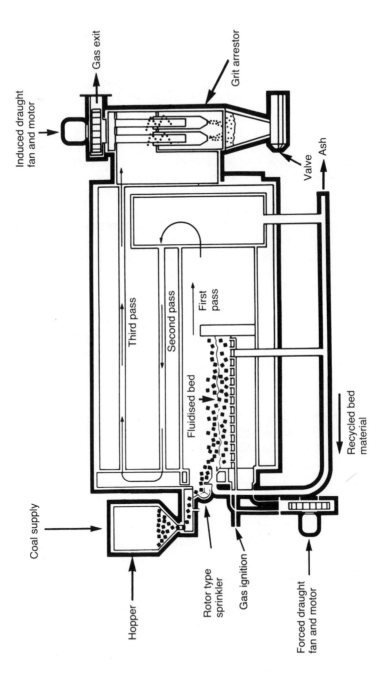

**14.17** Horizontal firetube boiler with fluidised bed

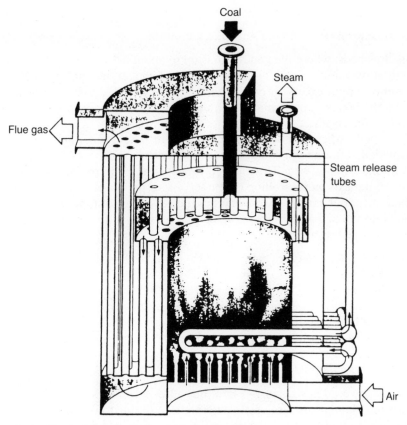

Coal

Steam

Flue gas

Steam release
tubes

Air

**14.18**   Vertical firetube boiler with fluidised bed
   (From British Coal)

suitable for accommodating fluidised beds than are horizontal firetube
furnaces. For modest steam outputs vertical boilers can be favoured (Fig.
14.18). The boiler output in this latter case is limited to about 4 MW for
two reasons.

1.   The furnace diameter, and therefore the heat input, is limited by Fig.
     3.1(6) of BS2790:1986 (see Fig. 12.1).
2.   The area of the water plane in a vertical drum is less than that in a
     horizontal drum of the same diameter and length >0.7854 diameter.
     This restricts the volume of dry steam that can be disengaged (see Ch.
     11). The steam velocity should not exceed about 0.055 m s$^{-1}$.

   The composite boiler, which is a water-tube furnace and a firetube
convection section (see Fig. 1.12), offers considerable advantages for
fluidised bed combustion. For large outputs, pressures and superheat,
however, full water-tube construction is necessary.

## 14.2.3.1 Types of fluidised beds

There are several types of fluidised beds, as follows.

1. Shallow beds, where the bed depth is up to about 0.3 m. In-bed tubes may or may not be used for controlling bed temperature. In their absence, either excess air or recirculated gases are used to extract heat from the bed and, particularly with the former, result in decreased boiler efficiencies.
2. Deep beds, where the bed depth is about 1 m. These generally provided better lateral mixing of fuel and air than shallow beds.
3. Rotating fluidised bed. This is a bed which is caused to rotate about a horizontal axis by controlling the underbed air pressures and quantities. For certain purposes it has the advantage of rapid mixing of fuel into the bed and overcoming the problem of distributing the fuel over a large bed area.

In both the former two cases bubbling occurs, but the bed surface is well defined at air velocities up to 2.5 m s$^{-1}$. Velocities above this figure increase the combustion rate, but the elutriation of bed material and partly burned fuel increase, needing early disengagement from the gases and return to the lower part of the bed for further burn-off. When this is done to a substantial extent the system is known as a 'circulating fluidised bed'.

## 14.2.3.2 Bed materials

Sand is mostly used in sizes of 1–2 mm, but other substances such as dolerite and alumina are also used. Where it is desirable to remove the sulphur compounds from the flue gases, suitably sized limestone can be incorporated in the bed. During operation bed materials degrade and are elutriated. The bed thus needs to be replenished from time to time or, with large beds, continuously. Most of the bed material is inert, only about 4% being fuel.

The fuel may contain extraneous material called 'stones'. These fall to the bottom of the bed and eventually interfere with fluidisation. They must be removed, either by an intermittent, manual process, or preferably by continuously removing material from the bottom of the bed, classifying it, returning undersized material to the bed and rejecting the oversized material.

Fluidised bed combustion is still in the developmental stage, but it has a number of advantages to offer.

1. The low combustion temperature, 900–950 °C, reduces the evolution of deposit forming material, e.g. soft ash and compounds of sodium, potassium and, with oil, vanadium and nickel.
2. A substantial amount of sulphur retention can be achieved by using limestone in the bed reduced in size to that of the bed material.

3. A wide variety of fuels, solid, liquid and gaseous, can be burned, at least in theory. It is therefore, again at present in theory, a universal burner.

4. The coefficient of heat transfer to in-bed tubes is very great. This means an economy of heated surface, and therefore of boiler size and cost.

There are however problems to be solved. From its nature a fluidised bed is a grinding machine, and carbon and inert material can be ground to very small ($\mu$m) sizes which are beyond the ability of cyclones to catch. Sophisticated gas clean-up systems, bag filters or electrostatic precipitators are therefore needed in order to comply with the emission regulations. The erosion of tubes, particularly in-bed tubes is another problem and, as has been mentioned earlier, so is the distribution of the fuel over and in the bed. So also, is the mechanised classification of bed material, with return or rejection according to size. It will take time and much effort to solve these problems.

## 14.3   Oil burners

The types of oil burner or atomiser in common use have been described in some detail in section 14.1.2 dealing with firetube boilers. The use of rotary cup burners on water-tube boilers tends to be restricted to the smaller sizes, but both pressure-jet and steam atomised types are widely used.

When the burner turndown achievable is insufficient to cater for the range of load variation on the associated boiler, on/off control can be adopted on a firetube boiler without undue operational difficulty. But such a system is normally only used on small hot-water boilers: modulated burners are mostly used on industrial boilers.

If an on/off system is applied to a water-tube steam generator, intolerable water level fluctuations occur due to the relatively small water capacity of the steam drum per unit of steam output compared with that of a firetube boiler. There may therefore be a desire to modulate the burners on a watertube boiler over a much wider load range than on a firetube boiler.

In addition to the various forms of pressure atomisation developed to give improved turndown, the ability to select the size and shape of the water-tube boiler furnace enables multiple burners to be fitted on all but small shop-assembled boilers (Fig. 2.8). It is therefore possible to obtain a higher turndown capability by taking burners out of service at reduced loads.

The inclusion of sufficient burner capacity enables each to be removed in sequence, and the atomiser to be cleaned while maintaining the boiler on load. The use of multiple burners necessitates extra attention being paid to the air distribution system, and burner windbox design, to ensure that each burner receives the correct amount of air, in uniform distribution, to efficiently burn the fuel put through it.

Removing burners from use introduces the need to keep the register and atomiser protected from overheating by radiant heat transfer from the furnace, a situation that does not arise on a multi-flue firetube boiler. The register is cooled by passing some of the combustion air through the non-operational burners. This results in an increase in the excess air supplied to the furnace to ensure that each burner receives sufficient for the oil it is burning (see section 4.1.2). The atomisers can be cooled by retracting them from the furnace, using tip shut-off devices which enable oil circulation to be maintained through the gun at all times, or by continuing to pass the atomising steam through the atomiser.

For turndown to be achieved automatically by these means necessitates a complex control system to carry out the functions required to achieve burner shut-off.

## 14.4   Gas burners

For firing rich gases, such as natural gas and refinery gas, on water-tube boilers, burners of the type described in section 14.1.3 are used. Rich gases are usually available at the burner at a pressure of about 2 bar. This, and the fact that with the use of multiple nozzles in each burner (Fig. 14.7), air and fuel mixing is very efficient, and reasonable turndown ratios are possible. As the gases are usually clean there is no necessity for spare burner capacity to ensure continuous full-load operation of the boiler.

Many industrial water-tube boilers fire low calorific value 'waste' gases (section 15.3). In such instances, the volume of gas to be handled is too great to enable the type of burners shown in Figs. 14.6 to 14.10 to be used. Burners for such fuels require air and gas passages of similar areas and typical burners have the combustion air passing through a central tube down which the auxiliary oil gun may be passed. The gas passes through an annulus around the central air tube. This arrangement gives reasonable air/fuel mixing for both fuels. Alternative arrangements have the burner divided into segments circumferentially with air and gas passing through alternate sections to give acceptable air/fuel mixing.

## 14.5   Other forms of firing

Alternative methods of firing boilers are continually being investigated. Some of the most publicised of these have been developed to enable boilers designed specifically for oil firing to be fired with coal during periods of uneconomical oil prices or during its non-availability.

### 14.5.1   Microfine coal

Coal reduced in size until 98% is below 40 $\mu$m is reported to burn with a flame characteristic similar to that of oil.[11] A high energy input to the pulverisers is required to achieve the desired coal fineness.

### 14.5.2   Coal/oil mixture (COM)

This was used as an interim measure to conserve oil consumption on boiler plant until such time that boilers specifically designed for firing coal would be purchased and installed. Mixtures containing up to about 50% by weight of coal can be fired.

### 14.5.3   Coal/water mixture (CWM)

This is an alternative to COM where no oil is used. Mixtures containing up to about 75% by weight of coal can be fired. The boiler efficiency will be significantly lower than for a coal-fired unit because of the heat loss to the high moisture content.

In the case of both COM and CWM, chemicals need to be added to maintain the coal in suspension, and the fuel handling systems are subjected to erosion by the coal. Due to the effects of the ash in the coal upon the fouling of boiler heating surfaces some reduction in output below that obtained with oil firing is inevitable.

The use of COM and CWM overcomes the problems on conversions associated with lack of space for fuel storage at the boiler front. Storage and handling facilities can be remote from the boiler as with oil firing and the fuel pumped direct to the burners on the boiler. The subject of making and burning coal/liquid mixture has been comprehensively discussed at recent international conferences.[12,13]

# References

1. Gunn, D. C. 'The effect of coal characteristics on boiler performance', *J. Inst. Fuel* 1952 (Mar.) p. 148.
2. MacDonald, E. J. and Murray, M. V. 'The effect of ash on the performance of shell boilers', *J. Inst. Fuel* 1963 (Jan.) p. 308.
3. MacDonald, E. J. and Murray, M. V. 'The effect of coal quality on the efficiency of a shell boiler equipped with a travelling-grate stoker'. *J. Inst. Fuel* 1955 (Oct) p. 479
4. Murray, M. V. 'Performance of an over feed coking stoker using 0.5 in – 0 smalls', *J. Inst. Fuel* 1957 (May) p. 276.
5. MacDonald, E. J. 'Smokeless combustion of 0.5 in smalls on coking stokers', *Engineering and Boilerhouse Review* 1952 (Nov.).
6. Fraser, R. P., Dombrowski, N. and Routley J. H. 'Performance characteristics of rotary cup burners', *J. Inst. Fuel* 1963 (Aug.) p. 316.
7. BS 799 1972: Part 4 *Atomising Burners over 36 litres/hour and Associated Equipment for Single and Multi-burner Installations.* Also 1979: Part 6 *Safety Times and Safety Control and Monitoring Devices for Atomising Oil Burner of the Monobloc Type.* British Standards Institution, London.
8. BS 5410 1978: Part 2 *Codes of Practice for Oil Firing Installation of 44 kW or Above Output Capacity for Space Heating, Water and Steam Supply Purposes.* British Standards Institution, London.
9. BS 5963:1981 *Specification for Electrically Operated Automatic Gas Shut-off Valves.* British Standards Institution, London.
10. BS 5885:1980 *Specification for Industrial Gas Burners of Input Rating 60 kW and Above.* British Standards Institution, London.
11. McDermid, B. and Falconer, R. D. *A Process Steam Boiler Conversion from Oil to Howden Microcoal Firing.* Inst. Mech. Eng. Conference on energy and the process industries 1985 (June).
12. Seventh International Symposium on Coal Slurry Fuels, Preparation and Utilisation. New Orleans, Louisiana 1985 (May).
13. Second European Conference on Coal Liquid Mixtures. Inst. Chem. Eng. EFCA Publication Series 44, London 1985 (Sept.).

# CHAPTER 15

# Waste fuel firing

## 15.0 Introduction

Wastes have been fired in boilers in a number of industries for many years both as a means of waste disposal and of energy recovery. Well-established examples are the firing of bagasse, which is the residue from sugar cane after the sugar has been removed, in sugar factories, and of blast furnace gas in steelworks. With increasing fuel and energy costs, consideration is now being given to energy recovery by the firing in boilers of any combustible wastes which can, in turn, reduce fuel and final product production costs.

The type of waste product being burned, and also the combustion process required to burn the waste, have a large influence upon the design of the boiler used. Particulate carryover, in the case of solid fuels, and corrosive properties of the products of combustion are significant considerations. As the less common wastes are used as fuels, it becomes necessary to carry out tests to determine the most suitable methods of firing these and a continuous process of boiler development is required.

The following chapter comments on some of the most common 'waste fuels', i.e. those other than the fossil fuels coal, oil and natural gas, and the influence these have on the boilers applied to their disposal. Waste fuels fall into one of the categories solid, liquid or gaseous, and combustion processes tend to be those used for the equivalent form of fossil fuel, or adapted versions of them. Typical analyses of the most common wastes fired are given in Tables 15.1, 15.2, 15.3 and 15.4.

## 15.1 Solid wastes

A large proportion of these are of fibrous or vegetable origin (often referred to as renewable fuels) having moisture contents that vary widely

**15.1** Typical fibrous fuel analyses – percentages by mass (ultimate)

| Constituent | Bagasse | Wood | Palm fruit waste | Rice husks | Straw[1] |
|---|---|---|---|---|---|
| Carbon | 24.72 | 25.2 | 36.7 | 40.75 | 36.0 |
| Hydrogen | 2.71 | 3.02 | 3.29 | 3.64 | 5.0 |
| Nitrogen | 0.42 | 0.12 | 0.98 | 1.13 | 0.5 |
| Sulphur | 0.11 | 0.05 | 0.28 | — | — |
| Oxygen | 20.64 | 20.61 | 24.86 | 30.08 | 38.0 |
| Ash | 1.40 | 1.0 | 3.89 | 15.4 | 4.75 |
| Moisture | 50.00 | 50.00 | 30.0 | 9.0 | 15.75 |
| Total | 100.00 | 100.0 | 100.0 | 100.0 | 100.0 |
| Gross calorific value (kJ kg$^{-1}$) | 9 473 | 10 000 | 13 962 | 15 190 | 12 560 |

**15.2** Typical analyses of untreated and pelletised municipal refuse – percentages by mass[2] (ultimate)

| Constituent | Untreated municipal refuse | Pelletised municipal refuse (RDF) |
|---|---|---|
| Carbon | 22.1 | 41.6 |
| Hydrogen | 1.7 | 5.4 |
| Sulphur | 0.2 | 0.4 |
| Nitrogen | 0.6 | 0.6 |
| Chlorine | 0.2 | 0.4 |
| Oxygen | 8.7 | 27.5 |
| Ash | 32.5 | 15.2 |
| Moisture | 24.8 | 8.9 |
| Metal | 9.2 | — |
| Total | 100.0 | 100.0 |
| Gross calorific value (kJ kg$^{-1}$) | 10 000 | 18 000 |

**15.3** Typical analyses of liquid wastes – percentages by mass

| Constituent | Black liquor solids |
|---|---|
| Carbon | 42.6 |
| Hydrogen | 3.6 |
| Sulphur | 3.6 |
| Oxygen | 31.7 |
| Nitrogen | 0.2 |
| Ash | 18.3 |
| Total | 100.0 |
| Gross calorific value (kJ kg$^{-1}$) | 15 352 |

**15.4** Typical gaseous fuel analyses – percentages by volume

| Constituent | Blast furnace gas | Catalytic regenerator (CO) gas | Off gas carbon black plant[3] | Coke oven gas |
|---|---|---|---|---|
| Carbon dioxide $CO_2$ | 12.64 | 10.56 | 2.39 | 2.5 |
| Carbon monoxide $CO$ | 22.33 | 7.54 | 6.90 | 7.0 |
| Hydrogen $H_2$ | 2.36 | — | 4.23 | 55.0 |
| Methane $CH_4$ | — | — | 0.16 | 27.0 |
| Acetylene $C_2H_2$ | — | — | 0.21 | — |
| Ethylene $C_2H_4$ | — | — | — | 2.7 |
| Argon $Ar_2$ | — | — | 0.37 | — |
| Nitrogen $N_2$ | 59.22 | 70.14 | 31.74 | 5.5 |
| Oxygen $O_2$ | — | 0.18 | — | 0.3 |
| Water vapour $H_2O$ | 3.45 | 11.58 | 54.00 | — |
| Total | 100.00 | 100.00 | 100.00 | 100.0 |
| Gross calorific value ($kJ\ m^{-3}$) | 3130 | 957.6 | 1602 | 19 000 |
| Gas temperature to boiler (°C) | Ambient | 600 | Ambient | Ambient |

from 10% to as high as 65% depending upon the source and the process to which they have been subjected on their way to the disposal point or boiler. Fibrous fuels also have a high volatile content, making them burn relatively easily once the moisture content has been reduced during the initial drying period in the furnace.

Other waste materials are those discharged by industry and the general public, for example municipal refuse and old vehicle tyres. Some of these 'fuels' may need significant pretreatment before they can be satisfactorily fired.

### 15.1.1   Fibrous fuels

Consideration is now being given to firing any vegetable waste having a reasonable calorific value, the system used depends very much upon the quantities available and the location of the installation with regard to the availability of labour for plant operation. A factor influencing the design of the plant is the form in which the fuel is fired, i.e whether it is prepared in a reduced size to enable it to be handled mechanically, or is left in large pieces necessitating some form of manual input.

Many of the vegetable wastes are seasonal and the supply is of very short duration, and this necessitaties long-term storage of these very low density and hence bulky materials. The use of an alternative fuel such as coal or oil may be necessary for a major part of the year, unless the steam demand is equally as seasonal, as in sugar factories which often only require steam when the sugar cane is being processed. An example of the short-term availability of a fibrous fuel is straw from cereal crops which is available at the time of harvesting for a period of about two weeks per year. It is a bulky material which, if it is to be used over a significant period of time will necessitate a large storage area, preferably covered.

Reclamation from storage of any fibrous fuel is difficult, as the material tends to bond together and will not, therefore, flow in the way of a granular material. It will often form a vertical surface (angle of repose 90° to the horizontal) and readily bridges over openings through which it is intended to flow. To recover such a material from storage will, therefore, require some mechanical means of picking it up or agitating it to enable it to be passed on to and handled by a conveying and feeding system. In the case of baled material such as straw the bales may have to be broken up before being fed into a furnace, although boilers are available which can handle whole bales.[4]

Material in large pieces or unprepared form can be fired on static or reciprocating grates, the latter agitating the fuel bed and improving combustion. Small or prepared fuel, i.e. less than 50 mm, can be fed into the furnace of a boiler using either mechanical or pneumatic distribution on to either static or moving grates. In such a system a quantity of the fuel will be burned in suspension, the amount depending upon the size of the fuel.

Most fibrous fuels have low ash contents and hence continuous ash discharge facilities are not always justified. Much of the ash content consists of potassium compounds which act as strong fluxes on refractory materials (see sections 15.1.1.3 and 10.3). The use of refractories in boilers should, therefore, be minimised by the use of water-cooled walls in water-tube boilers or, when this is not feasible, by using high-quality refractories containing 60% or more of alumina.

### 15.1.1.1 Bagasse

This is a low-density waste which must be disposed of. It has always been used as a fuel in cane sugar factories. It has a fibrous structure, a maximum dimension of about 50 mm and a moisture content of 45–55% as supplied to the boilers, see Table 15.1. A wide variety of combustion equipment has been used to fire this, some of the most common being pile burning in refractory furnaces (Fig. 15.1), firing on inclined or stepped grates and suspension firing over static, dumping or travelling grates (Fig. 15.2).

Refractory, cell or Dutch oven furnaces with stepped grates were widely used to fire early boilers, their use now is restricted to small outputs, up to about 6.5 kg s$^{-1}$ and low-technology areas. A 'Dutch oven' furnace is one installed outside the boiler as would be the case where a firetube boiler is

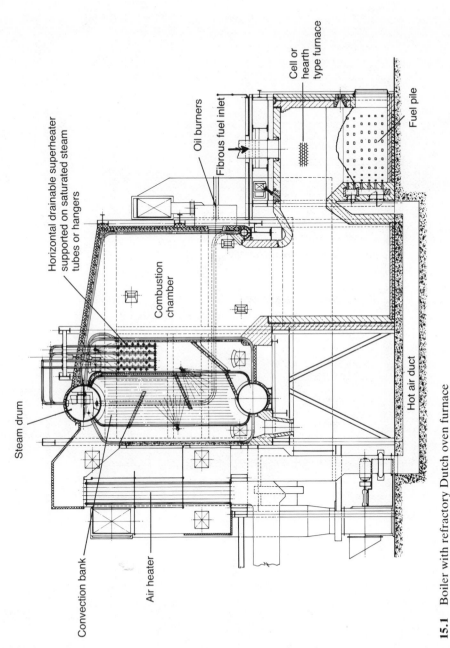

Horizontal drainable superheater supported on saturated steam tubes or hangers

Oil burners

Fibrous fuel inlet

Cell or hearth type furnace

Fuel pile

Combustion chamber

Steam drum

Convection bank

Air heater

Hot air duct

**15.1**  Boiler with refractory Dutch oven furnace

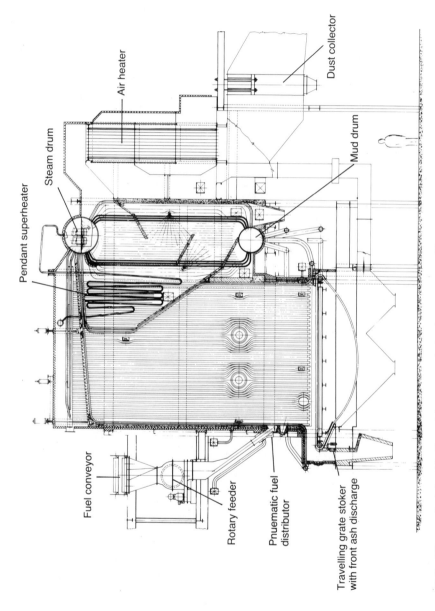

**15.2** Fibrous-fuel fired boiler with pneumatic feed on to a travelling grate

applied, or as illustrated in Fig. 15.1 for a watertube boiler. Refractory furnaces usually have to be shut down at intervals and the ash removed manually, although some have been fitted with a slow-moving travelling grate to give continuous ash discharge facilities, hence avoiding the inevitable reduction in boiler output while removing the ash. The stepped grate consists of an inclined grate made up of castings, with the fuel being fed in at the top, flowing downwards under gravity while burning, and leaving the ash at the bottom from where it can be removed manually with the boiler in operation.

Static 'pinhole' grates, consisting of perforated grate bars resting on boiler tubes,[5] and dumping grates, where the grate surface is tilted in sections to remove ash from the furnace, can be used for the whole range of outputs currently required. The fuel is distributed across the furnace by either mechanical (Fig. 14.15) or pneumatic distributors.

The relatively expensive travelling grate is usually only used where auxiliary coal firing is required, the continuous ash discharge facility it gives being necessary due to the relatively high ash content in coal compared with fibrous fuels.

The main features of boilers used for firing bagasse are low gas velocities to avoid tube erosion due to carryover of particulates, such as sand, and ample hoppers into which suspended matter from the gases can fall to avoid blockage of the gas passages.

Suspension fired boilers tend to have tall, moderately rated furnaces to reduce particulate carryover to a minimum. Nevertheless some form of dust collection equipment is necessary to avoid high induced-draught fan wear and chimney emission nuisance even though there may be no statutory requirements in the locality where the boiler is used. Until recently, thermal efficiency has been of no great concern, the main aim of the plant being to achieve a good balance between fuel availability and the energy or fuel required. Any surplus bagasse, being of low density, is expensive to transport for other uses and requires large areas for storage, although long-term storage is not recommended due to the fact that it will decompose and attract vermin.

Bagasse is a seasonal fuel and is hence only available during the cane processing season, i.e. when required as a fuel by the sugar mill. As the quality of bagasse can vary throughout the season there may be occasions when there is a surplus. To overcome this problem, 'variable efficiency' boilers have been installed to enable all the fuel to be disposed of as produced, except for a small stock retained for emergencies.[6] Other uses such as paper and board making are gradually being introduced with the result that increased thermal efficiency is required. Also, in some areas high-pressure boilers are used to increase electrical power generation, the surplus being exported to the local electrical grid.[7]

Bagasse drying is being introduced, where the heat in the flue gases leaving the boiler is used to drive off some of the moisture in the fuel. This has the effect of lowering the temperature of gas discharged to the chimney, giving a higher thermal efficiency than would otherwise be achieved, combustion is also improved.

Firetube boilers have now almost ceased to be used for burning bagasse because the bulk of the fuel necessitates using separate refractory furnaces, also the use of steam turbines to generate in-house power requirements requires steam conditions beyond their capability.

While most solid- or liquid-fuel fired boilers are fed with fuel from a bunker or tank, bagasse-fired boilers are unique in that the fuel is continuously conveyed direct from the mills where it is produced to the boilers, with no intermediate storage. This is done because of the difficulties already mentioned, associated with reclaiming fibrous fuels from storage. A breakdown in the milling plant which causes a stoppage in the flow of fuel will, therefore, result in a loss of boiler output. This is particularly significant with suspension firing unless a very efficient system of reclaiming from the fuel store is available or facilities for firing a readily available fuel such as oil or gas are fitted. Cell or hearth type furnaces, as illustrated in Fig. 15.1, in which the fuel burns in a pile, contain a store of fuel sufficient to provide boiler output for several minutes so they do not suffer from this problem.

There is a trend now, in small plants, for fibrous material to be pelletised. It has been found that with some materials, e.g. papyrus, which can be harvested as a fuel, an oil is exuded during compression. This binds the fibres together and also helps to make the pellet or briquette water resistant.[8]

### 15.1.1.2 Wood

Wood waste has long been used as a fuel in paper pulp mills where firing systems and boilers similar to those used for bagasse firing are applied. The moisture content can vary over a much wider range than that of bagasse, the actual value depending upon the method of transporting the logs to the mill, sometimes by river, and from where in the process the waste is derived. Moisture contents up to 65% have been known. This is approaching the limit at which combustion can be sustained without the use of a supplementary fuel. A typical analysis is given in Table 15.1. The timber industry tends to quote moisture contents as a percentage of the dry material content rather than as a percentage of the total. For example, a moisture content of 100% would mean 100% dry wood plus 100% water, i.e. 50% moisture in normal fuel terms. Care must, therefore, be taken to ensure that the basis of any moisture content quoted for wood is understood.

On large boilers, wood has to be prepared or hogged to a maximum dimension of about 50 mm to enable it to be handled mechanically and fed into the boiler.

In heavily forested equatorial areas, such as the Philippines, rapidly growing trees, (e.g. the Ipil Ipil), are cultivated as a fuel for small power generation plants (3 MW electrical) to supply electricity to small communities. With these there is a desire to fire the wood with minimum preparation, preferably in log form. A number of installations exist with this form of firing.[9]

Where the wood is harvested as a fuel the boilers will be designed to give higher efficiencies than so far used in cane sugar factories, since the wood is then a purpose-grown fuel and not a waste material that has to be disposed of.

### 15.1.1.3  Palm fruit waste

This is the fibrous residue from palm oil mills and consists of shell, kernel and fibre from the palm fruit from which the palm oil has been extracted (see Table 15.1). Palm oil mills are small compared with sugar mills, with a correspondingly low steam demand, up to about 3 kg steam per second being typical.

Firetube boilers, some externally fired, have been extensively used in the past, but the combination of uncooled refractory-lined furnaces and the low moisture content of the fuel resulted in heavy slagging of the furnace wall brickwork, and hence high maintenance costs. The trend over recent

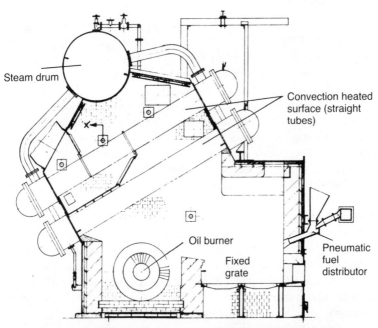

**15.3**  Water-tube boiler with straight tubes, fuel fed pneumatically on to a fixed grate
(From Towler Division, Babcock Robey Limited)

years has been to use simple low-technology water-tube boilers with some water cooling in the furnace, the fuel being fed into the furnace pneumatically and burned in suspension over fixed or dumping grates. A typical example is illustrated in Fig. 15.3.

### 15.1.2 Municipal refuse

This can be fired in mass as it is received from the collection vehicles. The process is currently known as incineration. No fuel preparation is carried out other than the crushing or breaking up of bulky items. Most incinerators are designed to handle items up to about 1 m diameter without pretreatment.

Refuse- or waste-derived fuels, RDF or WDF, are currently produced in two basic forms.[10] The first is as a floc, where combustible materials such as plastics and paper are separated from the waste and reduced to a maximum dimension of less than 50 mm.

The second is in the form of pellets, where the combustible content of the fuel is separated out and extruded through dies by presses to produce a product some 20 mm in diameter.

#### 15.1.2.1 Mass burning

Mass burning of crude or raw refuse is widely carried out, particularly in Europe where most of the combustion systems used were developed. There is an increasing interest in doing this both for energy conservation and due to the shortage of suitable refuse dumping areas for landfill. A typical analysis of a UK refuse is given in Table 15.2.

Municipal and industrial wastes have a very wide size distribution varying from 1 metre to dust. The fuel is therefore of a heterogeneous nature in that when it is fed on to a grate one section can be impervious to air flow with another section having sufficient voids to allow the free passage of combustion air, the net result being unbalanced combustion. This is particularly the case where the refuse is collected in plastic bags. All the combustion systems used are, therefore, inclined moving grates designed to agitate the fuel bed by reciprocating or rotating movement (Fig. 15.4) to enable the heat from the furnace to reach the incoming refuse to drive off excess moisture and to ensure that the combustion air reaches all the combustible material. Selection of the grate size for a given refuse throughput has to take into account the widely varying properties of the refuse, particularly the heating value. The boiler output is also subject to wide variation.

Boilers associated with incinerators have tended to be of the 'waste heat' type, where they were installed in ducting between the furnace and the chimney. Because this configuration has no heat-absorbing surfaces in the

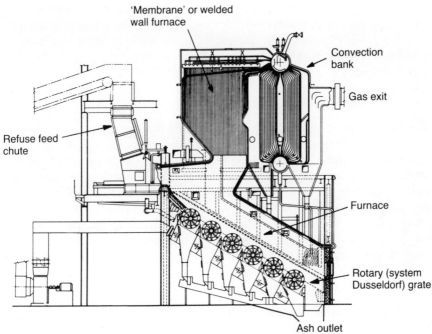

'Membrane' or welded
wall furnace

Convection
bank

Gas exit

Refuse feed
chute

Furnace

Rotary (system
Dusseldorf) grate

Ash outlet

**15.4**   Boiler fired with untreated refuse using a roller grate
(From Babcock Power Division of Babcock Energy Limited)

furnace, it is necessary to operate with high excess air quantities to achieve
acceptable gas temperatures to the boiler heating surfaces so as to prevent
slagging and high-temperature corrosion. While such designs still persist,
particularly for the smaller units of the pyrolytic type, the desire to recover
the maximum possible heat from the refuse, and to simplify furnace design
and reduce the quantity of brickwork and hence maintenance, has led to
use of water-cooled furnaces which enables the excess air to be reduced.
Figure 15.4 illustrates a typical incinerator/boiler unit for handling
municipal refuse.

Corrosion of the furnace water-walls has been experienced, apparently
due to the existence of a reducing atmosphere low down in the furnace.
This results from poor air-fuel mixing caused by the heterogeneous nature
of the refuse. To combat this, excess air values of about 100% are still
being used and the lower parts of the water-walls are covered with
refractory up to a point where satisfactory mixing of air and products of
combustion is achieved by the use of high-pressure secondary air jets.

Crude or untreated refuse and, indeed RDF/WDF, can contain
innumerable impurities not normally experienced in fossil fuels. These may
cause fouling and corrosion within the boiler if it is not suitably designed.
Many analyses can be obtained but, because the refuse properties vary so

widely, the relatively small quantities of impurities may not be evident since the source of particularly troublesome elements may not have been included in the sampling. It is, therefore, necessary when designing boilers associated with the combustion of refuse, to take note of all the data available from previous experience.[11]

### 15.1.2.2 Refuse derived fuel (RDF)

Floculated waste produced from refuse is fired in suspension, usually in conjunction with coal. Experience is varied throughout the world, but generally only some 25–50% of the heat input to the boiler is from waste.[12] 100% floc can be fired, but it is necessary to use a similar type of grate to that used for crude refuse, where the fuel bed is agitated. It will be necessary to have the ability to fire an alternative fuel to ensure security of steam output if floc is used as a boiler fuel in a process plant. The combustion system may, therefore, be a compromise for one or both of the fuels. The widespread use of floc may, therefore, be slow to develop.

### 15.1.2.3 Pelletised refuse

Refuse pellets, or densified refuse derived fuel (dRDF), are currently produced in relatively small quantities, about 20 000 tonnes per year, per plant[4], this being equivalent only to a continuous steam output of about 4 kg s$^{-1}$. The use of these is, therefore, likely to be limited to small firetube and water-tube boilers for heating purposes. The benefit of pelletisation, which requires considerable equipment and energy, is that the fuel density is much higher than crude refuse, thereby reducing transport costs and storage space. The size distribution is also greatly reduced, thus easing the problems referred to in section 15.1.2.1. The quality is reasonably consistent and pellets are easier to handle mechanically than are either floc or crude refuse. A typical analysis is shown in Table 15.2.

Pellets are fired on static grates, chain grate stokers and fluidised beds with varying degrees of success. Information is still being obtained with regard to possible ratings, emission and ash extraction systems.[4]

### 15.1.2.4 Hospital waste

This consists of dressings, parts of human bodies, packing materials, kitchen waste, etc., which must obviously be destroyed to sterile ash without smoke or odour. The incinerators for this purpose need to be of special design, using the principle of pyrolysis, and are the subject of BS3316.[13] They are refractory lined throughout and consist of a primary chamber in which the waste, supported on a fixed or moving grate is often burned with less than stoichiometric air, when it is known as starved-air combustion.

The products of combustion from this section, which are likely to be both smoky and smelly, are passed to a secondary chamber. In this latter the products of combustion, and further air, are brought into intimate contact. It may also be necessary to introduce an oil or gas flame to burn off combustible gases and smoke. The gases leave the incinerator at a temperature of about 1000 °C, which gives scope for considerable waste heat recovery in the form of steam or hot water to supplement the output from the hospital boiler plant.[14]

The boilers used for waste-heat recovery from hospital incinerators are firetube boilers, generally without a furnace tube, and are equipped only with a single pass of convection tubes of about 60 mm bore. These cool the gases to about 250 °C.

## 15.2   Liquid wastes

These are mainly effluents that cannot be discharged into drainage and sewage systems due to pollution hazards, although in some cases chemical recovery may also be a consideration in the method of disposal. Liquid wastes tend to be easier to handle than are solids but they may require some pretreatment, for example where the fluid has a high moisture content this may be reduced by evaporation to improve combustion characteristics. Acid tars have been burned in fluidised beds using coal as a support fuel.

### 15.2.1   Chemical recovery units

In wood pulp mills where the cellulose fibres in wood are extracted for paper-making, a liquid waste or liquor is produced which is a source of heat and power. This liquor has a solids content of 10–20%, a typical analysis of which is given on Table 15.3, it also contains chemicals used in the pulping process such as sodium sulphate or sodium sulphite. This effluent is called black liquor, or sulphite liquor, depending upon the chemical process being used. It is concentrated to about 68% solids using the products of combustion from the boiler to evaporate the water in direct-contact heat exchangers and is then fired in boilers to dispose of the combustible solids, recover the heat and to recover the process chemicals for reuse. The boilers used for this purpose are called chemical recovery units.

The liquor is sprayed on to the lower walls of the furnace in the boiler. The tubes in this area are given a protective layer of refractory to maintain the required gas temperature and to reduce tube corrosion. The combustibles burn off and produce gases containing a high loading of particulates of low fusion temperature. The process chemicals to be recovered are converted to a molten smelt, and run down the walls to the floor from where they are led out of the furnace to dissolving tanks for reprocessing.

Present-day power and steam requirements in pulp mills necessitate the use of high-pressure boilers of either the two-drum type with a single-pass convection bank or of the single-drum type (Fig. 2.5). Because of the high particulate loading and its low fusion temperature the furnace is tall to reduce the gas temperature to the convection surfaces. It is still necessary, however, for the convection bank, superheater and economiser to have very wide tube spacings to prevent bridging and blockage in the gas passages. The wide spacings also simplify slag and dust removal. Numerous sootblowers are included for on-load cleaning (see section 7.5).

## 15.3   Gaseous wastes

The most common of these are blast furnace gas produced in steelworks, CO gas from fluidised catalytic cracking plants and various rich refinery gases produced in oil refineries. These have been used as sources of energy in their respective industries for many years, and the combustion processes and boiler designs are well established.

### 15.3.1   Blast furnace gas (BFG)

Blast furnace gas has a low calorific value, the main combustible constituents being carbon monoxide and hydrogen (see Table 15.4). The gas is burnt in boilers to overcome disposal and pollution problems, in addition to recovering energy. It is first passed through regenerators or recuperators (stoves) to preheat the combustion air supplied to the blast furnace and then through gas scrubbers and precipitators to remove the high particulate content, it is, therefore, relatively clean, at atmospheric temperature and saturated with water vapour when supplied to the boiler. The gas pressure is only sufficient to overcome the resistance through the supply ducting and burners on the boiler (about 50 mbar g). The boiler must, therefore, operate under balanced-draught conditions.

Assuming that the supplementary or standby fuel for use in the event of non-availability of BFG is oil or natural gas, the boiler design is substantially the same as used for oil firing but with a moderate furnace volumetric rating of $0.25$ MW m$^{-3}$. While blast furnace gas will burn satisfactorily without support fuel, most BFG-fired boilers are fitted with permanent coke-oven gas pilots of 5–10% of the boiler full load heat input to ensure that safe conditions exist should there be a short-term deterioration in the quality of the fuel which could cause flame failure.

For a given boiler output the mass of the products of combustion from blast furnace gas is approximately twice that produced by oil firing, mainly because of the high inert content of the former. The boiler dimensions have to be carefully selected, therefore, if a high gas-side pressure drop is to be avoided.

Blast furnace gas, as most gases, burns with an almost non-luminous flame, giving a relatively low furnace heat absorption, (Table 15.5), nevertheless the high gas mass flow from the furnace results in a temperature of the gas leaving the furnace some 120 °C below that obtained when firing oil at the same steam output.

**15.5**   Typical furnace/boiler convection surface heat distribution

|  | Percentage of heat transfered to steam | |
|  | Blast furnace gas | oil |
| --- | --- | --- |
| Furnace | 22.4 | 52.5 |
| Convection surfaces | 77.6 | 47.5 |

The large difference in the mass of products of combustion for BFG and oil has a significant influence upon the design of the superheater to ensure that the desired steam temperature can be achieved with both fuels. The conditions for oil firing establish the superheater heating surface required, the performance with BFG firing necessitates steam temperature control equipment with a high capability to dispose of the excess superheat transferred to the steam (section 2.5.1). The design of airheaters for use with BFG is also affected, the gas mass to air mass ratio being very high. The reduction in gas temperature that can be achieved with a given rise of air temperature is small compared with that with an oil-fired unit (see Table 7.1). The induced-draught fan is affected too. The performance required when firing oil is sufficiently low to create a control (turn-down) problem (section 7.2.5). Two-speed fans are often used to improve the situation.

### 15.3.2   CO gas from fluidised catalytic crackers (FCC)

The gases produced are the result of continuous catalyst regeneration by substoichiometric pressurised combustion of the coke deposited in catalytic cracking plants in oil refineries. The low excess air and high pressure (about 2.7 bar) are necessary to protect the catalyst from damage due to overheating. The low calorific value gases are discharged with a carbon monoxide content of 7–8% (Table 15.4), and a temperature of approximately 600 °C. They therefore contain both recoverable sensible heat and combustion heat in the ratio of approximately 44% to 56%, depending upon the flue gas temperature selected. The carbon monoxide must of course be burnt off before the gases can be discharged to atmosphere.

Both conventional water-tube type boilers and separate uncooled combustion chambers followed by heat recovery surfaces are used to recover the heat contained in the gases. Because of the high gas temperature to the boiler the burners used to mix the gas and air have to be constructed from special materials.

To ensure combustion stability and boiler safety supplementary fuel is permanently fired into the furnace to produce a stabilising flame, the heat input from this source being some 10% of the total. The quantity of gas passing through the boiler is very high for a given output compared with firing only oil and hence these boilers usually operate under pressurised conditions making use of the high gas pressure available at the outlet from the FCC unit to overcome the pressure drop across the boiler.

The types of boiler used are similar to those employed for oil and gas firing, but as there can be a significant carry-over of catalyst dust from the FCC unit the boiler design and cleaning facilities must take this into account.

### 15.3.3   Refinery gas

This term covers the rich gaseous by-products from the production of fuel oils from crude oil. They usually have a high calorific value on a volume basis, i.e. higher than does natural gas, since they consist mainly of the high molecular mass hydrocarbons. There is normally no problem with their combustion and, as they are clean fuels, no special provisions are required with regard to boiler gas-side cleaning.

Boilers supplied to oil refineries have facilities to burn a variety of fuels to enable them to utilise those for which a surplus may exist at any time. These can include liquid fuels in addition to the refinery gases. These alternative fuels, which can include asphalt and pitch, will have a greater influence on the design of the boilers than will the gases.

## References

1. *Fuels and Combustion Handbook* (1st edn). McGraw-Hill, p. 128.
2. Hufton, P. F. *Refuse derived Fuel – prospects for the Industrial User*    *I.E. and Inst. Chem. Eng.* 1985 (Feb.).
3. Hurley, E. G. *Burning Low Calorific Value Gases* (circa 1960).

4. *Energy Management – FOCUS Issue 3*, 1985, Department of Energy.
5. Levy, P. W. and Kenny, D. 'The watercooled stationary grate'. *Proceedings of Australian Society of Sugar Cane Technologists* 1984.
6. Levy, P. W. 'The air cooled condenser system of efficiency variation for bagasse fired boilers', *Proceedings of Australian Society of Sugar Cane Technologists* 1981.
7. Eller, W. M. 'Power generation in sugar factories', *Combustion* 1974 (Sept.).
8. Joseph, S. *Biomass*, CEA Annual Conference, 1986.
9. Duckworth, P. A. 'Rural woodburning power stations' Inst. Mech. Eng. Seminar *Package Power Stations for Export 1985*.
10. Porteous, Dr A. *Supplementary fuels from Municipal Solid Waste*. Combustion Engineering Association, Manchester, 1983
11. Krause, H. H., Vaughan, D. A. *et al*. A series of papers 'Corrosion and deposits of solid waste' published by *ASME Journal for Engineering for Power* **95** No. 1 1973 (Jan.), **96** No. 3 1974 (July), **97** No. 3 1975 (July), **98** No. 3 1976 (July), **99** No. 3 1977 (July), **101** 1979 (Oct.).
12. Fiscus, D. E., Peterson, R. D. *et al*. *RDF Cofiring in the Electricity Utility Industry* (US), ASME Solid Waste Processing Conference Nashville 1983 (June).
13. BS 3316: 1973 *Specification for Large Incinerators for the Destruction of Hospital Waste*. British Standards Institution, London.
14. Priest, G. M. *Producing Steam from the Combustion of Solid Waste*. Inst. Mech. Eng. Seminar on Energy Recovery from Refuse. 1985.

CHAPTER 16

# Waste-heat boilers and thermal storage

## 16.0 Introduction

Waste heat can be recovered from a high-temperature source using any convenient low-temperature fluid, which may be one used in a process, for example the feedstock which may require to be heated or combustion air for fuel used in the process. This chapter refers only to steam or hot-water generators and for this purpose a waste-heat boiler is defined as one in which heat is transferred from hot gas or other heat source that has been produced for a purpose other than the generation of steam or hot water.

## 16.1 General discussion

Boilers have for many years been fitted to industrial furnaces in steelworks, and used in metallurgical processes and glass making to recover heat from the high-temperature flue gases produced by the melting process which would otherwise have been discharged to atmosphere. In the early days, heat recovery would only have been carried out if there was a requirement for steam or hot water. More recently, increased fuel and energy costs have created a need for economy, and waste-heat boilers have been designed to operate at high temperatures and pressures, thus enabling electric power generation by steam turbines to be carried out.

Restrictions applied to chimney emissions have made it necessary to cool waste gases to a temperature acceptable to dust collection equipment. Waste-heat boilers are a convenient way of doing this, with power generation being incorporated to at least provide the in-house power requirements even where exporting surplus power is not encouraged.[1] Both the turbine exhaust and any steam surplus to requirements in such cases can be condensed and reused for boiler feedwater.

Waste-heat boilers are widely used in conjunction with chemical processes, where their main purpose is to cool hot gas produced by the reactions or heating of the feedstock before it passes to a subsequent part of the process for further chemical reactions to take place. Examples are sulphuric acid plants where pure sulphur or iron pyrites is burned to produce oxides of sulphur[2], or the steam reforming furnace where a fuel is burned to heat the feedstock by indirect contact, thus producing both hot flue gases and hot process gases.[3]

The steam produced may be used to drive compressors incorporated in the process or exported to a factory main at a suitable pressure, as is frequently done with large chemical and refinery complexes. Some of the steam may be used in the process itself, as with the steam reformer.[3]

With many waste-heat boiler schemes, and particularly those used in chemical processes for gas temperature control, the amount of steam generated or of heat transferred is controlled by the quantity of heat available in the gases and not by the demand for steam from the boiler. In some processes with a high steam demand it may become necessary to incorporate an auxiliary-fired boiler or some form of firing within the waste heat-boiler system. Where the gas temperature on leaving a waste-heat boiler is critical to the downstream process, some form of gas outlet temperature control will be required. This can take the form of a gas bypass (see Fig. 16.1), where some of the gas is diverted round some or all of the heating surface, giving a stream of high-temperature gas which is then returned to the main low-temperature gas stream leaving the boiler,

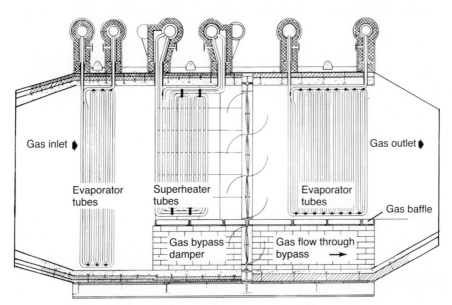

**16.1**  Forced-circulation water-tube boiler for sulphuric acid process

giving a higher mixed temperature. For such a system to operate, the boiler has to be oversize, so that when all the gas is passed through the boiler the gas temperature on leaving is below that desired.

This philosophy of oversizing should always be applied to waste-heat boilers where the performance cannot be exactly predicted, on the basis that it is always easier to reduce the performance of a boiler (raise the gas leaving temperature) than it is to increase it. The quantity of gas bypassing the heating surface will be controlled by some form of damper or valve, and it is essential that the gas temperature at the controlling device is such that a practical design of damper or valve that will give good reliability is possible. Where high boiler gas inlet temperatures exist, an acceptable environment for the controlling device can be achieved by a number of methods depending upon the boiler design used. An example as applied to water-tube boilers is to reduce the gas temperature by passing all the gas through part of the heating surfaces before dividing the flow into main stream and bypass as illustrated in Fig. 16.1. When using firetube boilers, two can be arranged in series on the gas side with the controlling device arranged in the second unit.

Where a wide turndown of throughput is required, two boilers in parallel on the gas side can be used, one giving a high gas outlet temperature and the other a low one to give gas temperature control when the two streams are mixed.

Waste-heat boiler systems can, and often do, include economisers and superheaters which may be adjacent to the boiler or placed remotely in other parts of the process (see Fig. 16.2). Where both steam and gas temperature control are required, the control system can become quite complex. Superheaters or economisers can also be used to control process gas temperatures. The desired control is achieved by bypassing steam, in the case of superheaters, or water, in economisers, around the heating surface. Flow variation is by a system of bypass valves controlled from the gas outlet temperature (see Fig. 16.3).

Air preheaters can only be used if there is a process requirement for hot air to burn a fuel or feedstock.

In waste-heat boilers used to recover heat from low-temperature waste gases, for example gas turbine exhaust, the amount of heat that can be recovered, the maximum steam temperature and the maximum working pressure at which the boiler can operate are dictated by the ability to maintain a temperature difference between the gases and the heating surfaces rather than by the desire to achieve a selected final gas temperature. In such cases the controlling temperature difference is that at the outlet of the evaporator. This is known as the 'pinch point' and is illustrated in Fig. 16.4. Minimum temperature differences of about 20 °C can be economically justified depending, of course, upon current costs.

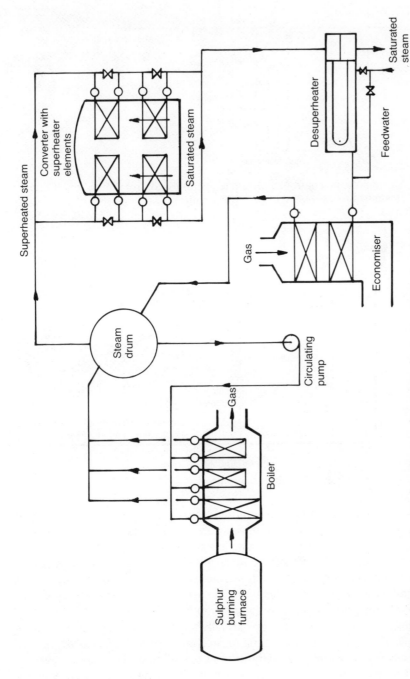

**16.2** Typical steam and water flow diagram for a waste heat recovery system on a sulphuric acid process

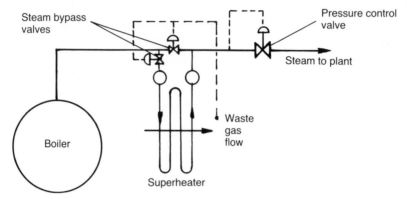

**16.3** Steam pressure control and steam temperature control using a superheater steam bypass as applied to waste-heat boilers

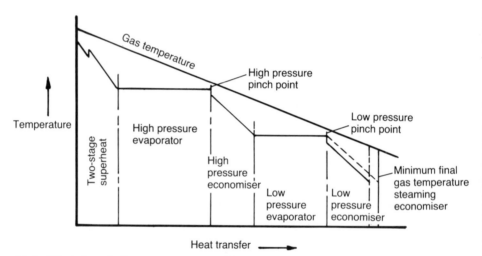

**16.4** Waste-heat boiler temperature profile

To overcome the limitations imposed by the pinch point, increased heat recovery can be achieved from the waste gases by installing evaporators in series on the gas side. These operate at various steam pressures, reducing as the gas temperature falls (Fig. 16.4). The low-pressure steam can be used for feedwater heating in a de-aerator, for process work if a convenient demand exists, or for space heating. The maximum output is achieved by using steaming economisers as indicated in Fig. 16.4.

Thermal efficiency is not a term usually applied to waste-heat boilers because the boiler maker has little or no control over the gas flow, its composition or the inlet temperature to the boiler. Where a specific gas outlet temperature is a design requirement, this is more important than the quantity of heat recovered.

Where the steam produced is exported to other plants, it may be necessary, in order to maintain a constant pressure and hence water temperature within the waste-heat boilers, to install a pressure control valve in the steam main leaving the boiler (Fig. 16.3). This ensures that the temperature of the water in the boiler is constant, thus maintaining its performance. The valve is arranged to be controlled from the pressure on the upstream side, a rise in pressure above the set point opening the valve to allow more steam to flow out of the boiler. A reduction in the upstream pressure would have the reverse effect. Variations of the downstream pressure produced by off-plant equipment would initially cause a variation in flow through the valve and hence in the upstream pressure, until the variation was of such a magnitude that the valve control responded to correct it. Such a valve is often known as a 'surplus' or 'overflow' valve and is a valuable adjunct to any boiler, fired or otherwise, where it is desirable to maintain a constant boiler pressure irrespective of demand.

Waste-heat boilers can be designed to operate under either positive or negative pressure conditions. Those recovering heat from flue gases are usually fitted with induced-draught fans and hence operate under a negative pressure. Exceptions are boilers recovering heat from gas turbines and diesel engines. Process waste-heat boilers usually operate under a significant positive pressure.

The available gas-side pressure drop is often a major factor for establishing the dimensions of a waste-heat boiler, particularly one operating with clean gas. A high allowable pressure drop gives minimum heating surface, minimum enclosure dimensions and hence minimum cost.

## 16.2   Boiler types

Both firetube and water-tube boilers can be used for waste-heat recovery applications. As with fired boilers, the firetube boiler will usually be the least costly and should be used wherever it can be suitably applied. Figure
  Figure 16.5 gives an indication of the amount of heat that can be recovered from waste gases.

### 16.2.1   Firetube boilers

Firetube boilers used for waste-heat recovery are rarely fitted with large-diameter furnaces since heat transfer is mostly by convection with some non-luminous radiation. Small-bore tubes are more effective under these circumstances (see section 5.4). Furnaces can be fitted to reduce the gas temperature before it enters the smaller diameter tubes should this be necessary to avoid tube blocking by molten suspended matter in the gases. Firetube boilers are ideally suited for processes where the gases to be

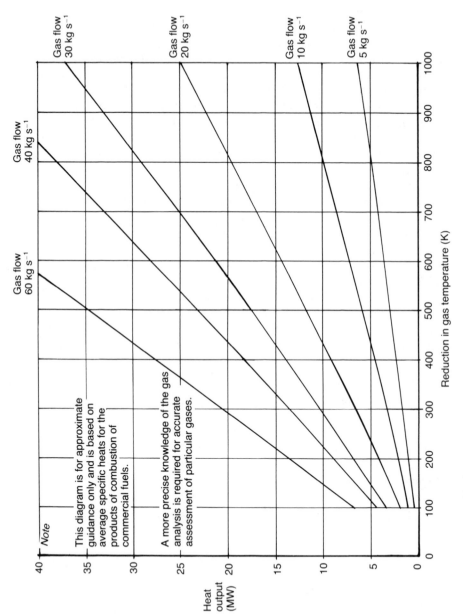

**16.5** Heat recoverable from waste gases

cooled are at high pressure, since they form a natural pressure vessel with the high-pressure gas in the tubes.

When used for high steam pressures, above 30 bar, the firetube shell will be completely filled with tubes so spaced as to give the minimum shell diameter and also to give the necessary support to the tube plate, thus eliminating the large areas unsupported by tubes into which stay bars have to be fitted. This also means that the shell thickness can be minimised (see section 1.2.). In such cases the boiler is fitted with a separate steam drum in which the steam and water separation takes place. The drum can be supported from the firetube boiler by risers and downcomers of the required mechanical strength (Fig. 16.6), or separately supported and connected to the firetube boiler by circulating pipework of the desired size and flexibility. One steam drum can be used for multiple boilers (Fig. 16.7), with the obvious economy of mountings and instrumentation.

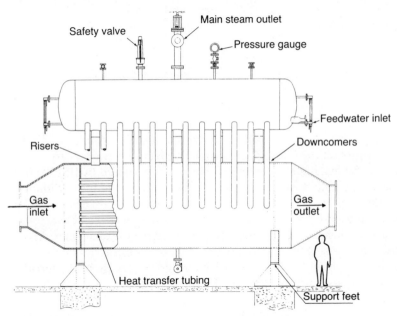

**16.6** Firetube waste-heat boiler with separate steam drum

Firetube boilers having no furnaces or steam space can be used for steam pressures up to 100 bar, but the design requires rigorous stress analysis because of the thick shells and tube plates required. When operating with high gas pressures and full of small bore tubes they are capable of outputs far in excess of those achieved from fired firetube type boilers.

When using high gas pressures consideration has to be given in the design of the tubes and tube plates to stresses occurring in the event of

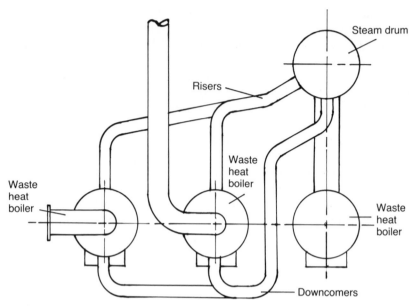

**16.7** Multiple firetube waste-heat boilers with a single steam drum

there being a high gas pressure but no steam pressure. To do this, it will be necessary to refer to BS5500.[4]

If the gas inlet temperature is high, above say 1000 °C, it may be necessary to protect the inlet tube plate with a layer of refractory to comply with the maximum tube plate temperature required by BS2790 (see particularly Appendix C of that Standard). The ends of the tubes within the tube plate can be protected by the insertion of ferrules. These are steel or ceramic tubes inserted into the boiler tube, leaving a space between the two or having a layer of insulation between to reduce the heat transferred to the tube as it passes through the tube plate. The result is a tube metal temperature significantly below that which would exist without the ferrules, but the draught loss or pressure drop through the tubes is increased by their use.

High gas pressures allowing high pressure drops and hence high gas mass velocities to be used, enable very high heat-transfer coefficients to be obtained (Ch. 5). To prevent the tubes overheating due to steam blanketing occurring with high heat fluxes, care must be taken with the tube layout of large-diameter firetube boilers to ensure that there is adequate water circulation.

In instances where the temperature of gas leaving the boiler is low, 200–300 °C, firetube boilers having vertical tubes with upward gas flow have been used and the output of the boiler is controlled by varying the water level in the boiler. The heat-transfer coefficient from the tube to the steam in the steam space is lower than that to the water (Ch. 5), therefore lowering the water level in the boiler reduces the effectiveness of the heating surface and hence the boiler output. This principle has been used extensively for boilers recovering heat from the exhaust gases of diesel engines.

To control the output of water-tube boilers in similar applications, sections of the heating surfaces have been fitted with valves at the inlet of the water-side. These can then be isolated as desired to reduce the effective surface area and hence output.

Firetube boilers, when used with gases containing the high particulate loadings experienced with some processes, can be subjected to severe fouling. This results in the blockage of the tubes, which can then be difficult to clear[5] (see Ch. 10). Tube-erosion can also be a problem if high gas velocities are used to optimise the heating surface. Depending upon the flow patterns, the rate of erosion can be proportional to (gas velocity)$^{3.5}$, see reference 6.

## 16.2.2 Water-tube boilers

Water-tube boilers used for waste-heat applications can be of the natural (or forced circulation) or once-through types using single or multiple drums as required. The limitations on working pressure imposed by the various types are practically the same as for fired boilers (see Ch. 2), and there is no reasonable limit to the output for which a water-tube waste-heat boiler can be designed other than that imposed by the heat available in the waste gases.

The configurations of water-tube boilers used for waste-heat applications are probably more varied than those for fired installations. They are particularly suited to applications recovering heat from dust-laden gases because of the flexibility of tube diameters, spacings and arrangements (vertical or horizontal) available. Cleaning devices such as sootblowers and rapping gear can easily be accommodated.

Forced circulation enables the boiler heating surfaces to be arranged in the manner most suited to the process, often avoiding the use of extensive high-temperature ductwork that may be necessary to transport the gas to a natural circulation boiler using vertical tubes. Forced circulation also enables the steam drum to be placed remote from the gases, in the position most convenient for the plant layout and to be used to cater for a number of boilers.

For high gas pressure applications, for example nitric acid plants where gas pressures of the order of 8 bar are used, a circular pressure containment vessel is required for the heated surfaces. The use of forced circulation allows the heated surfaces to be arranged to suit the shape of the vessel Fig. 16.8.

Forced-circulation boilers are also considered to be more suitable for rapid start-up due to the fact that pumped circulation gives a more uniform

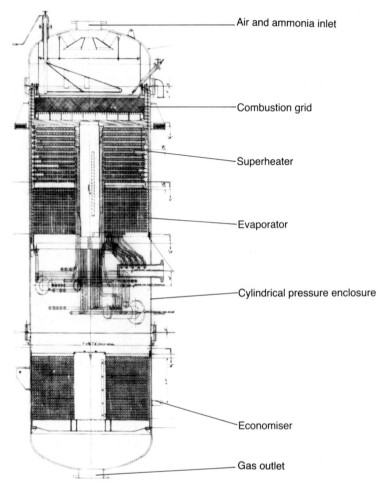

Air and ammonia inlet

Combustion grid

Superheater

Evaporator

Cylindrical pressure enclosure

Economiser

Gas outlet

**16.8** Forced-circulation water-tube boiler for nitric acid process (high gas pressure)

rate of heating from cold and, the single drum, being independent of rigid tube connections between drums as with a two-drum arrangement, is subjected to lower stress set up by the differential expansions.

Firetube waste-heat boilers can be arranged with forced circulation. This enables a number of widely dispersed boilers to be arranged to use one steam drum.

## 16.3   Combined schemes

It is quite common for both firetube and water-tube boilers to be used in combination on one plant. An example of the versatility of boilers when used in conjunction with a chemical process is illustrated in Fig. 16.9, the

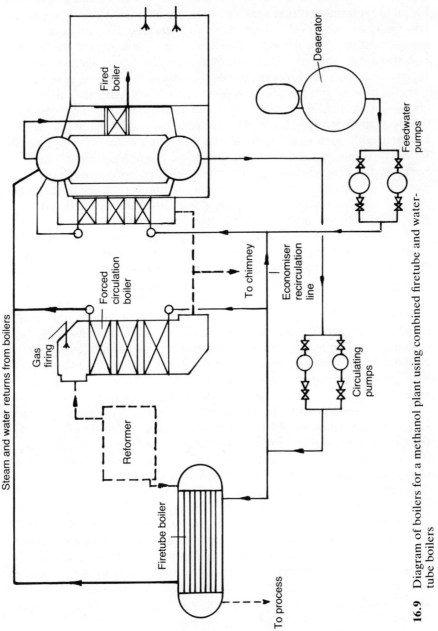

**16.9**  Diagram of boilers for a methanol plant using combined firetube and water-tube boilers

diagram for a methanol plant. With this example the pressurised process gas from the reforming furnace is passed through a firetube boiler for gas temperature control. The flue gases produced by the combustion of the fuel in the reforming furnace are passed through a forced-circulation water-tube boiler arranged with vertical downward gas flow. As a process requirement supplementary gas firing is included at the inlet to the flue gas boiler to reduce the oxygen content of the flue gas. In this instance the steam demand for the process exceeds that available by recovery of heat from the process and flue gases. A conventional two-drum natural-circulation gas-fired water-tube boiler is included in the system to generate the shortfall in output. This fired boiler includes a superheater and economiser through which pass the total steam and water flow respectively for the three boilers. Circulating water for the firetube and water-tube waste-heat boilers is taken from the fired boiler, pumped through the waste-heat units, and the steam and water mixture is returned to the steam drum of the fired boiler for separation along with that generated in the fired boiler. The steam then passes to the superheater.

The criteria for the selection of tube spacings and metal temperatures that influence the design of fired boilers also apply to waste-heat boilers. The requirements can, however, be more onerous due to the wide range of process gases with which they come into contact. It may be necessary to use stainless steel or other special materials to combat low-temperature corrosion, even though this may only be in the form of a protection to prevent the occurrence of a major disaster in the event of a tube leak.

Process gas containing high percentages of hydrogen requires special consideration. When in contact with materials at high metal temperatures hydrogen embrittlement can occur unless the correct materials are selected. Guidance on this is given by the 'Nelson Curves'.[7]

When the gases passing through waste-heat boilers are absolutely clean, very close tube spacings and extended-surface tubing with closely spaced fins can be used. This is particularly advantageous where the gas is at high pressure, necessitating a pressure vessel to enclose the boiler surfaces. The use of closely spaced tubes and extended surfaces minimises the size, thickness and cost of the enclosure.

An example of closely spaced plain tubing is the coiled surfaces used with nitric acid processes where the gas pressure is high (Fig. 16.8). The process gas is produced by the exothermic reaction between ammonia and air over a catalyst enclosed at the boiler inlet and contains no particulate matter. The spaces between the tubes are restricted only by the gas-side pressure drop and the practical limitations of the supporting and manufacturing tolerances of the heating surfaces. Spacings between tubes down to 5 mm have been used.

## 16.4    Gases containing molten particles

The gas from metallurgical furnaces is noted for containing molten particulates which solidify on cooling, particularly when they make contact with the relatively cool boiler heating surfaces. Typical of these is the copper smelter, which produces gas with a heavy loading of copper particles. Heat-recovery boilers for such applications have a large radiant water-cooled chamber preceding the relatively closely spaced convection surfaces (Fig. 16.10). This chamber is designed to reduce the temperature of the gas and particulates below the softening temperature of the latter. This, in combination with a low gas velocity which allows precipitation of some of the particulates, reduces considerably the fouling within the convection surfaces. The radiant chamber water-cooled walls are arranged with ample facilities for on-load cleaning by sootblowers or by manually slicing through the suitably positioned and shaped access apertures through the welded wall enclosure.

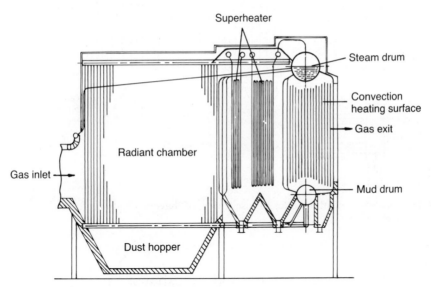

**16.10**   Natural-circulation boiler for a copper smelter

## 16.5    Gases containing erosive dust

Where the gas contains very high loadings of particulates, for example sulphuric acid plants where the gas is produced from heating iron pyrites in a fluidised bed, water-tube boilers having horizontal gas flow across vertical or pendant heated surfaces are used. The tubes can easily be cleaned by long retractable sootblowers or by mechanical rapping gear. Deposits removed from the tubes, and those particulates settling out under gravity, fall into a series of hoppers beneath the tubes from where they can be removed by suitable grit handling equipment. This design of boiler is particularly suitable where the particulates are highly abrasive. Velocities below 15 m s$^{-1}$ will minimise tube erosion. The boilers can be of forced- or natural-circulation type depending upon the formation of the tube banks. Figure 16.11 illustrates a forced-circulation boiler used for this purpose.

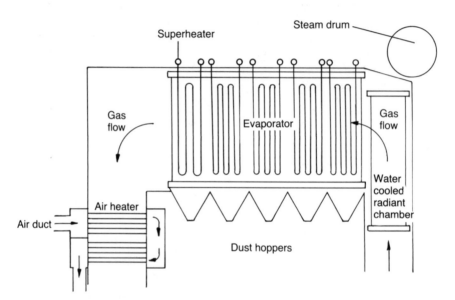

**16.11**   Forced-circulation boiler for use with gases having a high particulate content

## 16.6   Diesel engine and gas turbine exhaust

Waste-heat recovery from the exhaust gas produced by diesel engines and gas turbines involves the handling of relatively large quantities of gas at moderate to low temperatures. Typical values are a large gas turbine of 85 MW electrical output producing 350 kg s$^{-1}$ of gas at 515 °C and a 21 MW diesel engine producing 42 kg s$^{-1}$ of gas at 320 °C.

Firetube boilers are widely used for the smaller units, and the steam conditions generated by the larger units may well be within the capability of firetube boilers, but the limits of boiler size imposed by the manufacturing facilities and transport, combined with the larger quantities of gas involved, would necessitate a large number of units. Ducting costs and space requirements make firetube boilers uneconomical under such circumstances compared with a single water-tube boiler, particularly if the latter uses extended surface tubing.

The gas-side pressure drop across boilers for diesel and gas turbine exhaust heat-recovery applications is critical, since any back pressure imposed by the boiler reduces the engine output. Induced-draught fans are not used because of the cost, the power required because of the large gas volume involved and the problem of control. Boiler gas-side pressure drops of 15 to 25 mbar are used.

Systems utilising the heat in the exhaust gas from diesel engines and gas turbines are referred to as 'combined cycles'[8], where the steam generated is passed through a steam turbine for the generation of additional power or, as 'combined heat and power' (CHP) systems, where the steam is used as a heat source for process or heating.

Supplementary firing is often incorporated in these boilers where the exhaust gas, having a high oxygen content (15–16% by weight), is used to supply the oxygen for the combustion of the additional fuel. The majority of the heat losses are already accounted for by the exhaust gases and hence the efficiency of combustion of and heat recovery from the supplementary fuel is very much higher than in a conventional fired boiler. Such an arrangement is used in a combined cycle where additional power is generated, and very good plant efficiency is achieved[8] compared with that with a conventional fired power plant. The introduction of supplementary firing adds heat to the gases and hence increases the gas temperature. This enables higher steam pressures and temperatures to be achieved, in addition to the increased output.

The boiler arrangement can vary from one with fuel burners in the gas inlet duct before the convection heating surfaces in the case of a small addition of heat and hence small temperature rise, to one with a conventional water-cooled furnace when high fuel inputs are required, giving high gas temperatures. In some cases, it may be necessary to generate the normal full load steam output when the gas turbine is not in operation, giving the gas temperatures experienced in a conventional fired boiler.

The selection of plain or extended-surface tubing for use in the convection heating surfaces depends upon the properties of the fuel used in the gas turbine and for supplementary firing. Clean fuels such as gas or light distillates enable finned tubing to be used. Figure 16.12 illustrates a supplementary fired extended-surface natural-circulation water-tube boiler as applied to heat recovery from gas turbine exhaust gas.

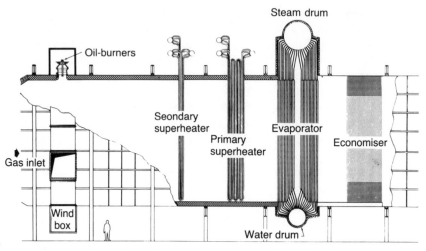

**16.12** Natural-circulation boiler using extended-surface tubes for gas turbine exhaust application

## 16.7 Availability and reliability

Where a waste-heat boiler is incorporated into a process, it is important that it does not impair the availability of that process. That is, it must not reduce the period of operation of the process between consecutive normal periods of shut-down for maintenance, or reduce the output below that which would be experienced by the process on its own. It should be appreciated that the cost of loss of process output for several hours can far exceed any maintenance or repair costs associated with a boiler failure. Boiler reliability is therefore a major factor in the plant design.[9] With particularly difficult gases a multiplicity of boilers may be favoured to give a measure of standby. Alternatively, regular replacement of part or all of the heated surfaces may be a solution.

With waste-heat recovery it is absolutely essential that the boiler designer has made available to him full information regarding the analysis of the gases and any particulates contained in them. Even small quantities of impurities can be significant under the conditions that exist in some units. The results of any previous experience are very important, particularly where fouling, tube erosion or corrosion have occurred.

As will be seen, the design and application of waste-heat recovery equipment are very diverse. There are a number of other common processes, such as basic oxygen blown converters in steelworks, where special design features have developed to suit the peculiarities of the process. Comprehensive coverage is not possible within this book.

## 16.8   Thermal storage and thermal storage boilers

### 16.8.1   Thermal storage (steam accumulator)

The storage of heat to meet peak load conditions has been practised for
many years. Heat is stored as saturated water in a separate vessel
downstream of the boiler, in which case the vessel is known as a 'steam
accumulator'. The use of an accumulator to store and supply steam relies
upon the fact that the enthalpy of saturated water increases with increasing
saturation pressure (see reference 10).

During periods of low steam demand the accumulator is charged by the
boiler, using the excess steaming capacity of the latter. By doing so the
pressure of the accumulator is raised to or near to that of the boiler. When
a heavy demand for steam occurs, in excess of the full load capability of
the boiler, steam is obtained from the accumulator by allowing the
pressure in the accumulator to fall. This reduces the enthalpy required by
the saturated water. The excess enthalpy is absorbed as latent heat and
water is evaporated, i.e. steam is 'flashed off'.

Such a system requires that the boilers operate at a pressure above the
operating pressure of the process to which the accumulator supplies steam.
It does, however, enable the total operating boiler capacity to be equal to
the average total steam demand rather than to the maximum, which may
be considerably in excess of the average even though only for a short
period of time.

The principle of operation and the similar process which occurs when
high-pressure blowdown is discharged to atmosphere, are indicated by the
following example.

Process steam is required at 7 bar gauge and the boilers operate at and
charge the accumulator to 15 bar gauge

| | |
|---|---|
| Enthalpy of saturated water at 15 bar gauge | $= 844.7 \text{ kJ kg}^{-1}$ |
| Enthalpy of saturated water at 7 bar gauge | $= 697.1 \text{ kJ kg}^{-1}$ |
| Enthalpy released when the pressure in the accumulator falls from 15 to 7 bar gauge $= 844.7 - 697.1$ | $= 147.6 \text{ kJ kg}^{-1}$ |
| Enthalpy to generate saturated steam from saturated water at 7 bar gauge | $= 2064.9 \text{ kJ kg}^{-1}$ |
| Assuming the steam reaches the process at 7 bar gauge and saturated, the steam released by the accumulator $= \dfrac{147.6}{2064.9}$ | $= 0.0715 \text{ kg per kg of water}$ |

The water capacity of the accumulator is calculated by determining the quantity of water required to give the desired quantity of steam over the required time period. Because the available drop in pressure in an accumulator is usually small, a large volume of water is required to generate the desired steam flow, thus requiring a large containment vessel. The accumulator is charged by discharging the steam from the boiler into the water through a number of nozzles near the bottom of the vessel which is usually a horizontal cylinder to give the desired steam release area.

The system is illustrated diagrammatically by Fig. 16.13. Further information on the subject can be obtained from reference 10.

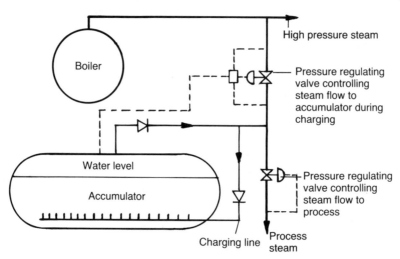

**16.13**   Diagrammatic arrangement of a boiler and steam accumulator system

### 16.8.2   Feedwater accumulator

An alternative method of heat storage is to use excess steam during periods of low demand to preheat the feedwater in a pressure vessel on the delivery side of the feed pump. When a heavy demand on the boiler occurs, the feed is already at saturation rather than at normal feedwater temperature, thus enabling the heat transmitted to the boiler to be used for evaporating rather than for heating the water.

This latter system, known as the 'feedwater accumulator', has the disadvantage of increasing the steam disengagement rate in the boiler and may cause priming to occur. Nevertheless, it is used where peak loads are modest and within the design criteria of the boiler. Its advantages are that a vertical storage vessel can be used. This requires less floor space than the horizontal vessel used for the steam accumulator, and the system operates at constant pressure.

## 16.8.3  Thermal storage boiler

In the early 1950s, Dr E. G. Ritchie, formerly Chief Engineer of the Ruth's Company, and then Director of the Engineering Division of the British Coal Utilisation Research Association, invented the 'thermal storage boiler' which achieved thermal storage without the need for an additional pressure vessel. The principle of the thermal storage boiler is similar to that of the feedwater accumulator. When peak steam flows in excess of the maximum continuous output for which the boiler is designed are required, the feedwater valve is closed, thus preventing the relatively cold feedwater being fed to the boiler. To generate steam under these conditions, where all the water contained in the boiler is at saturation temperature, only the enthalpy of evaporation has to be added to the water instead of the additional sensible heat necessary to raise the water from feedwater temperature to saturation temperature. For the same heat input to the furnace this can increase the short-term output by as much as 40% where cold feedwater is used.

For example, if steam is being raised at 15 bar absolute from feedwater at 80 °C:

| | |
|---|---|
| The enthalpy of saturated steam is | 2789.9 kJ kg$^{-1}$ |
| The enthalpy of saturated water is | 844.7 kJ kg$^{-1}$ |
| The enthalpy of feedwater at 80 °C is | 335 kJ kg$^{-1}$ |
| The enthalpy required to raise steam from feedwater at 80 °C = 2789.9 − 335 | = 2454.9 kJ kg$^{-1}$ |
| The enthalpy required to raise steam from saturated water is 2789.9 − 844.7 | = 1945.2 kJ kg$^{-1}$ |

Steam generated from saturated water is therefore $\dfrac{2454.9}{1945.2}$ times that from feedwater at 80°C  =  1.26,

i.e. 26% more than from feedwater at 80 °C for the same heat input to the boiler. The increase obtainable is reduced with higher feedwater temperatures.

The volume of water held in the boiler must, of course, be sufficient to provide the water evaporated over the period during which the feedwater valve is closed, without allowing the water level to fall to a dangerous level. This necessitates a larger diameter boiler than would otherwise be required, giving a greater water coverage over the tubes at normal operating water level. A very long gauge glass (about 1 m long) and special water level controls are needed. Care must also be taken to ensure that the steam disengagement rate is not excessive. During periods of low steam demand the water level is allowed to reach a preset high level; when a heavy demand occurs this level is allowed to fall without the feed pump operating. When a predetermined low level is reached the feed pump operates normally under the influence of normal water level controls.

# References

1. Horton, R. W. *Experience with Steam and Power Generation from Municipal Refuse*. Inst. Mech. Eng. Seminar on energy recovery from refuse incineration. 1985 (Feb.).
2. Duecker, W. W. and West, J. R. *The Manufacture of Sulphuric acid*. Reinhold.
3. Lyon, S. D. *Development of the Modern Ammonia Industry*. Tenth Brotherton Memorial Lecture, April 1985 (Apr.).
4. BS 5500: *Unfired Fusion Welded Pressure Vessels*. British Standards Institution, London.
5. Hall, G. S. and Knowles, M. J. D. *Good Design and Operation Reaps Benefits on UK Refuse Incineration Boilers Over the Last 10 years*. Inst. Mech. Eng. Seminar on energy recovery from refuse incineration. 1985 (Feb.).
6. Moir, M. K. and Mason, V. 'Tube wear in sugar mill boilers', *Proceedings of Australian Society of Sugar Cane Technologists*. 1982.
7. Bonner, W. A. 'Preview of new Nelson curves', *Hydrocarbon Processing* 1977 (May) p. 165–7. Also API Publication 941 (2nd edn) 1977.
8. Mitchell, J. E. *Combined Cycle Generation*. BEAMA power conference, Singapore, 1974 (Nov.).
9. Hinchley, P. 'The engineering of reliability into waste heat boiler systems.' I. Mech. E. proceedings 1979 Vol. 193 No. 8.
10. Goodall, P. M. *The Efficient use of Steam* I.P.C. Science and Technology Press, **18**.

# Index